## STOCKHOLM INTERNATIONAL PEACE RESEARCH INSTITUTE

SIPRI is an independent international institute dedicated to research into conflict, armaments, arms control and disarmament. Established in 1966, SIPRI provides data, analysis and recommendations, based on open sources, to policymakers, researchers, media and the interested public.

The Governing Board is not responsible for the views expressed in the publications of the Institute.

## GOVERNING BOARD

Ambassador Sven-Olof Petersson, Chairman (Sweden)
Dr Dewi Fortuna Anwar (Indonesia)
Dr Vladimir Baranovsky (Russia)
Ambassador Lakhdar Brahimi (Algeria)
Jayantha Dhanapala (Sri Lanka)
Ambassador Wolfgang Ischinger (Germany)
Professor Mary Kaldor (United Kingdom)
The Director

## DIRECTOR

Dan Smith (United Kingdom)

**STOCKHOLM INTERNATIONAL PEACE RESEARCH INSTITUTE**
Signalistgatan 9
SE-169 70 Solna, Sweden
Telephone: +46 8 655 97 00
Fax: +46 8 655 97 33
Email: sipri@sipri.org
Internet: www.sipri.org

# The New Arctic Governance

SIPRI Research Report No. 25

Edited by
Linda Jakobson and Neil Melvin

OXFORD UNIVERSITY PRESS
2016

# OXFORD
UNIVERSITY PRESS

Great Clarendon Street, Oxford OX2 6DP
United Kingdom

Oxford University Press is a department of the University of Oxford.
It furthers the University's objective of excellence in research, scholarship,
and education by publishing worldwide.
Oxford is a registered trade mark of Oxford University Press
in the UK and in certain other countries

© SIPRI 2016

All rights reserved. No part of this publication may be reproduced, stored
in a retrieval system, or transmitted, in any form or by any means, without the
prior permission in writing of SIPRI, or as expressly permitted by law, or under
terms agreed with the appropriate reprographics rights organizations. Enquiries
concerning reproduction outside the scope of the above should be sent to
SIPRI, Signalistgatan 9, SE-169 70 Solna, Sweden

You must not circulate this book in any other form
and you must impose the same condition on any acquirer

British Library Cataloguing in Publication Data
Data available

Library of Congress Cataloging in Publication Data
Data available

Typeset and originated by SIPRI
Printed in Great Britain by
Clays Ltd, St Ives plc

Links to third party websites are provided by Oxford in good faith and
for information only. Oxford disclaims any responsibility for the materials
contained in any third party website referenced in this work.

This book is also available in electronic format at
http://books.sipri.org/

ISBN 978-0-19-874733-8

# Contents

| | |
|---|---|
| **Preface** | viii |
| **Acknowledgements** | x |
| **Abbreviations** | xi |

**1. Introduction** — 1
*Neil Melvin and Kristofer Bergh*

    I. The new Arctic governance — 1
   II. Key themes of the new Arctic governance — 4
  III. The structure of the report — 11
Figure 1.1. The Arctic region — 3

**2. Security in the Arctic: definitions, challenges and solutions** — 13
*Alyson J. K. Bailes*

    I. Introduction — 13
   II. Dimensions of Arctic security: hard security and conflict — 15
  III. Environment, energy and economics — 23
  IV. Civil emergencies — 32
   V. Societal and human security — 34
  VI. Multidimensional security and Arctic governance — 37

**3. Understanding national approaches to security in the Arctic** — 41
*Kristofer Bergh and Ekaterina Klimenko*

    I. Introduction — 41
   II. North America — 42
  III. Russia — 48
  IV. The Nordic countries — 59
   V. An emerging circumpolar security architecture — 66
  VI. Conclusions — 73

## 4. Russia's Arctic governance policies   76
*Andrei Zagorski*

    I. Introduction   76
    II. The law of the sea   77
    III. The continental shelf   81
    IV. Navigation   87
    V. International fisheries   95
    VI. Security   99
    VII. The Arctic Council   104
    VIII. Conclusions   107
Figure 4.1. Norway's CLCS claims, the continental shelf outside of the EEZ   85
Figure 4.2. Russia's CLCS claim, 2015   86
Figure 4.3. NEAFC and international waters in the Arctic Ocean overlap   97

## 5. North East Asia eyes the Arctic   111
*Linda Jakobson and Seong-Hyon Lee*

    I. Introduction   111
    II. China's Arctic activities and policies   113
    III. Japan's Arctic activities and policies   127
    IV. South Korea's Arctic activities and policies   135
    V. Conclusions   143
Figure 5.1. Arctic sea routes and potential resources   112
Figure 5.2. Comparison of the Northwest Passage and the Northern Sea Route   113

## 6. The Arctic Council in Arctic governance: the significance of the Oil Spill Agreement   147
*Svein Vigeland Rottem*

    I. Introduction   147
    II. A brief history of the Arctic Council   148
    III. The Oil Spill Agreement   158
    IV. The Arctic Council in Arctic governance   166
    V. Concluding remarks   172

| Figure 6.1. | Arctic Council structure | 151 |
| Table 6.1. | Arctic Council membership | 158 |

**7. Conclusions**    174
    I. In summary    174
    II. Observations and implications for future Arctic governance    181

**About the authors**    190

**Index**    193

# Preface

Climate change is dramatically transforming the Arctic. In 2015 the region experienced record air temperatures, combined with a new low in peak ice extent. The annual average air temperature was 1.3°C above the long-term average (1981–2010) and was the highest since modern records began in 1900. In some parts of the Arctic, the temperature exceeded 3°C above the average. The Arctic Ocean also reached its peak ice cover for the season on 25 February—15 days earlier than the long-term average and with the lowest extent since records began in 1979.

These dramatic changes are creating severe problems for the indigenous populations of the Arctic, who are often dependent on natural resources for their survival, as well as for other communities that live and operate in the region. While it is hoped that the agreement to limit climate change reached at the 2015 Paris Climate Summit will slow global warming, there is still likely to be a 4°–5°C increase in the average Arctic temperature by 2050.

In these conditions, adaptability will be critical to survival. So too will be shared and effective governance in order to create the political consensus to act and the capacities to respond effectively. As this SIPRI Research Report highlights, over the past two and half decades the Arctic has experienced the emergence of a unique system of governance driven by regional challenges, notably climate change, and based on scientific knowledge about the region.

Pivoting around the Arctic Council, the formal and informal mechanisms of governance in the region have brought together the Arctic states, and increasingly other actors within and outside the region (notably China), to form crucial new agreements on the foundation stone of international law and to promote a spirit of Arctic cooperation.

Security cooperation is an important dimension of building these new approaches to managing and developing the Arctic. In recent years, important progress has been achieved on 'soft security' questions in the region, while 'hard security' has begun to be addressed in informal ways. This has helped to move the Arctic away from the militarization it experienced during the cold war and to fashion the agreements necessary for regional security as it experiences the impacts of climatic shifts.

However, as this report underlines, the continuing success of Arctic governance cannot be guaranteed. It rests, fundamentally, on cooperation among the Arctic states. Yet with the Arctic increasingly affected by globalization, future economic development will likely depend on actors and markets located far from the region. Arctic governance will therefore need to evolve continually in order to meet the new conditions in and around the region.

A possible deterioration in the international security environment would represent one of the most significant challenges to the spirit of cooperation. From a security perspective, the Arctic does not constitute a single space but is fragmented by linkages to wider security groupings that stretch far beyond the region. For this reason, the confrontation between Russia and the Euro–Atlantic community, which has spiralled outwards from the Ukraine crisis, has also affected security cooperation in the Arctic.

Despite these difficulties, the Arctic Council (chaired by the United States in 2015–17) has continued to function and new agreements have been reached, notably the Arctic Coast Guard Agreement of October 2015. The conviction among Arctic states that cooperation in the region must continue, even while they fundamentally disagree about developments elsewhere, suggests that Arctic governance relations have acquired a significant legitimacy and durability. A strong sense of stewardship and a conviction that it is critical to build shared rules suggests that multilateralism remains alive and well in the Arctic.

At the same time, the crisis in relations between Russia and the Euro–Atlantic community points to the need for further trust building, transparency and understanding of security issues in the Arctic. As this report makes clear, rebuilding and enhancing security cooperation in the region is a priority, not just for managing military tensions but also for maintaining and developing the regional governance that is so critical for the Arctic's future.

Dan Smith<br>
SIPRI Director<br>
Stockholm, December 2015

# Acknowledgments

We would like to offer sincere thanks to the authors of the individual chapters and their patience with the production process of the book. On behalf of the authors, we also thank their many colleagues and contacts around the world who helped with their research. We are also indebted to the SIPRI editors of this volume, Dr Ian Davis and Annika Salisbury, who did a marvellous job in editing the text. Finally, we want to thank the Swedish Foundation for Strategic Environmental Research (Mistra) for its financial support for this project.

Linda Jakobson and Dr Neil Melvin
Sydney and Stockholm, December 2015

# Abbreviations

| | |
|---|---|
| A5 | Arctic Five |
| AC | Arctic Council |
| ACIA | Arctic Climate Impact Assessment |
| AEC | Arctic Economic Council |
| BEAC | Barents Euro-Arctic Council |
| CHOD | Chiefs of Defence |
| CLCS | Commission on the Limits of the Continental Shelf |
| COSCO | China Ocean Shipping (Group) Company |
| CSBM | Confidence- and security-building measure |
| EEZ | Exclusive economic zone |
| EU | European Union |
| IMO | International Maritime Organization |
| KOPRI | Korea Polar Research Institute |
| LNG | Liquefied natural gas |
| MARPOL | International Convention for the Prevention of Pollution from Ships |
| NATO | North Atlantic Treaty Organization |
| NEAFC | North East Atlantic Fisheries Commission |
| NORAD | North American Aerospace Defense Command |
| NORDEFCO | Nordic Defence Cooperation |
| NGO | Non-governmental organization |
| NSR | Northern Sea Route |
| OSCE | Organization for Security and Co-operation |
| RFMO | Regional fisheries management organization |
| SAR | Search and rescue |
| SAO | Senior Arctic Official |
| SOA | State Oceanic Administration (China) |
| SOLAS | International Convention for the Safety of Life at Sea |
| UNCLOS | United Nations Convention on the Law of the Sea |
| USEUCOM | United States European Command |

# 1. Introduction

NEIL MELVIN AND KRISTOFER BERGH

## I. The new Arctic governance

Just a few years ago journalists and experts were predicting that the Arctic would be a location of conflict over resources in the future. Today, observers identify the region (see figure 1.1) as a key zone of cooperation, built on an emerging 'stability architecture' linked to the Arctic Council (AC) and based on international law. In this context the Arctic states have been able to foster a new Arctic spirit and even to incorporate a variety of non-Arctic states, notably from northern Asia as well as Europe, on the sidelines of the emerging structures of Arctic governance.

The contemporary form of Arctic governance thus differs fundamentally from that which existed during the cold war. This new order rests not on military strength, but on common interests and a willingness to pursue them. While a fundamental shift from the era of East–West confrontation, the appearance of this 'new' form of Arctic governance is not—as the authors in this volume highlight—the result of revolution but rather evolution over the past 25 years.

The rise of new forms of governance in the Arctic has been unexpected. During the cold war, the military confrontation between the Soviet Union and the transatlantic community extended into the Arctic: the region was a zone of competition and hostile relations. Today, the daunting challenge of climate change has transformed perceptions, bringing together all states involved in the Arctic around science and environmental protection. The potential for resource extraction and new transportation routes has further encouraged cooperation in the region.

The future development of the Arctic is not, however, without challenges. A key issue will be the degree to which security issues affecting the region can be managed effectively. Despite making important progress in building cooperative forms of governance in the Arctic, security cooperation remains underdeveloped. The increasing interest in the Arctic has led to a securitization of the region as states seek to consolidate their effective sovereignty.

## 2 THE NEW ARCTIC GOVERNANCE

Although the creation of new capabilities falls short of a remilitarization of the Arctic to the levels seen during the cold war era, potential confrontation in the region remains a risk and mutual suspicions linger. The region continues to be militarily significant because of the strategic nuclear forces operating on submarines in the Arctic. Moreover, as the conflict in Ukraine has highlighted, the positive achievements in Arctic relations of recent years remain vulnerable to geostrategic confrontation among the Arctic players in other theatres.

The central aim of this report is to explore and highlight the emerging patterns of governance in the Arctic through an examination of an interlinked set of questions.

1. What constitutes the current form of Arctic governance?
2. What explains the emergence of this form of governance in the Arctic?
3. What are the future challenges to Arctic governance?
4. Does the experience of building multilateral, cooperative and peaceful governance in the Arctic offer lessons to other parts of the world?

The report does not aim to provide a comprehensive review of the governance structures and institutions that influence the Arctic, but rather to (*a*) illustrate the central aspects of Arctic governance, (*b*) identify trends within it, and (*c*) highlight the key dynamics and actors that affect governance in the region. A central theme is to identify how the issue of security is affecting the forms of governance in the region.

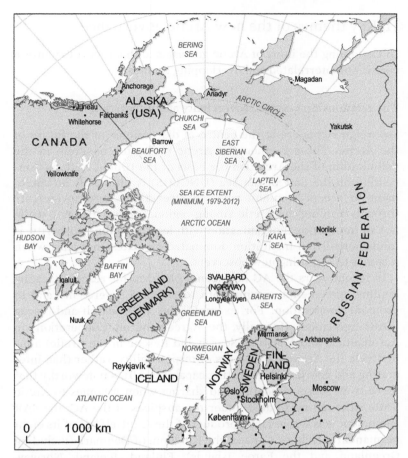

**Figure 1.1.** The Arctic region
*Credit*: Hugo Ahlenius, Nordpil, <https://nordpil.se/>.
*Source*: National Snow & Ice Data Center, 'Monthly Sea Ice Extent', 19 Nov. 2015, <https://nsidc.org/data/seaice_index/>.

## II. Key themes of the new Arctic governance

In examining the issue of Arctic governance this report explores a core set of interrelated and overlapping themes.

### The relationship between sovereignty and integration

Sovereign rights are at the centre of discussions on the future of the Arctic. The agreement by Arctic states to cooperate in the delimitation of the Arctic on the basis of the United Nations Convention on the Law of the Sea (UNCLOS) is ultimately an agreement to accept common rules to strengthen state sovereignty through dividing the Arctic among regional states. The AC, while not an international organization with legally binding authority, is fundamentally an institution dominated by states—not least when it comes to accepting new observers.

In recent years the AC has been a platform for reaching legally binding agreements, serving both to enhance sovereignty and to promote cooperation. For example, the 2011 search-and-rescue (SAR) agreement (formally the Agreement on Cooperation on Aeronautical and Maritime Search and Rescue in the Arctic) aims to divide the Arctic into sectors of responsibility, rather than integrating and coordinating existing capabilities. It is noteworthy that there is no sense of pooling and sharing when it comes to Arctic SAR. Although negotiated under the auspices of the AC, the SAR agreement was not in fact a product of the AC itself but of its eight permanent members: Canada, the Kingdom of Denmark (including Greenland and the Faroe Islands), Finland, Iceland, Norway, Russia, Sweden and the United States. The same applies to the 2013 Agreement on Cooperation on Marine Oil Pollution Preparedness and Response in the Arctic.

Therefore, the primary interests of the Arctic states have been to (*a*) consolidate control over their territories, (*b*) extend continental shelf limits, and (*c*) increase influence. Rather than compete, the Arctic states have chosen to cooperate on the basis of international law to achieve these aims. Thus, at present there is little basis for major conflict, with the only territorial dispute being the relatively minor one over Hans Island. This is a dispute between Canada and Denmark (on behalf of the Greenland self-government) and multilateral cooperation offers the best way to resolve it. The positive atmosphere in the region thus rests, to a significant degree, on the agreement of all the Arctic states to cooperate in the shared aim of dividing the Arctic cake and determining limited rules of the game for future exploitation of the region.

*Good fences may make good neighbours, but what will the continuing focus on sovereignty mean for the future of Arctic governance? Can Arctic governance develop further without forms of integration and pooled sovereignty in the region? Are there real limits on steps in this direction? Does working together come more naturally to the smaller Arctic states and less so to the Arctic superpowers, Canada and Russia?*

**The Arctic states and globalization**

Globalization has changed the way that the state is seen. Flows of people, capital and ideas are blurring traditional boundaries and challenging the sovereignty of states; climate, conflicts and crime rarely consider the borders on a map. Thus, responding to such issues requires new forms of international cooperation. The Arctic does not exist in a vacuum and it is subject to the forces of globalization just as any other part of the world. Today's Arctic is a region facing several borderless challenges, yet discussions about its future are framed in a traditional view that places the role, interests and capabilities of the state at its centre.

While Arctic states remain the dominant actors in Arctic governance, there has been a realization that the region is connected to the international order and that, to a significant degree, its successful development will require the involvement of non-Arctic interests (notably Asian states and markets). This has promoted the evolution of Arctic governance and the acceptance of slightly more inclusive—but still highly limited—approaches. The extent to which international interests can be accommodated with those of the Arctic states, however, remains a central question.

The decision to grant non-Arctic states permanent observer status in the AC (from the previous ad hoc observer status) at the 2013 Kiruna ministerial meeting suggests that the relationship between international and Arctic politics is at the beginning of a complex process, which is likely to require careful negotiation and balancing. At the same time, a group of international non-state actors, notably including Greenpeace, was not granted observer status. While scientists and, to some degree, indigenous peoples have found a place within Arctic governance, more overtly political groupings remain outside many of the governance formats. As the attention given to Arctic issues grows, the activities of human rights groups, environmental groups and other non-governmental organizations (NGOs) are likely to increase in the region.

Commercial interests are becoming increasingly prominent in the Arctic, challenging the previous state-centric focus. Ultimately, the business sector will drive the development of the region and this is likely to require different types of governance—including, perhaps, corporate governance and responsibility. Any significant economic development of the Arctic will require cooperation with outside actors. For example, Russian oil and gas will need to be exploited in cooperation with non-Russian companies outside of Russia that have advanced technology and the capital necessary for the substantial investments required to operate in the region. The Northern Sea Route (NSR) will also only become commercially viable through significant engagement and investment from beyond the region.

*How will non-Arctic states, non-state actors and commercial groups shape Arctic governance in the future? Can the existing state-centric forms of Arctic governance accommodate the rise of new actors and interests in the region? And if so, how? Will new forms of governance be needed?*

## Public versus private actors

One of the most important actors driving development in the Arctic is the business sector. The prospect of expanding resource extraction, shipping and tourism, among others, in an increasingly open Arctic is a powerful motivator for presence in the region. The Arctic's commercial potential is also one of the key drivers of national policies towards the region. This is especially clear in Arctic littoral states, where economic development of the region is high on the political agenda.

Business interests in a region as challenging as the Arctic do not develop autonomously, independent of states. This raises important questions about the role of public versus private interests in the region. Perhaps in the future the most important actors in the Arctic will not be Canada, Denmark, Norway, Russia and the USA, but rather Shell, BP, Rosneft, Gazprom, Statoil and commercial shipping firms.

There are, however, different approaches to the market and its operation among the various Arctic stakeholders. In some cases, such as Russia and China, it can be difficult to distinguish between public and private interests, raising questions about which actor is actually driving policy and what interests are shaping particular actions. Concepts of national champions—whereby large organizations are expected not only to seek profit but also to 'advance the interests of the nation'—and resistance to 'foreign' involvement are also likely to influence the role that commercial interests can play in the region.

*What will be the relationship between states and private actors in the development of the Arctic? What is the appropriate way to govern private actors in the Arctic: the market, state regulation, multilateral agreements or international law? Are the existing Arctic governance arrangements sufficient or are new ones necessary to promote and regulate businesses?*

## Regional multilateralism and its sustainability in the Arctic

Most observers today agree that the Arctic states are pursuing their interests in the Arctic in a mostly non-confrontational way and that the multilateral frameworks created to address challenges in the region are working well. The AC is increasingly attractive as a forum and it may eventually even emerge as an international organization. These developments point to a thickening of regional institutions and a growing sense of the Arctic as a distinct region for the interaction of a group of states. Yet it is not a regionalism defined by a movement towards political integration or ideas of shared sovereignty, nor a political ambition in this direction.

Against the background of previous concerns about the likelihood of conflict in the Arctic, the positive example of multilateral cooperation in the region is interesting, especially in a time when regional multilateralism more generally seems to be in crisis. However, close analysis of the basis for cooperation that underpins multilateralism in the Arctic highlights that the agreement among states to cooperate (in the face of international claims on the Arctic) rests on the division of the region among themselves. This limited foundation for cooperative governance suggests that the current situation of strengthening multilateralism could be a transitional phenomenon. The future of Arctic governance may well involve a shift in the region's variable geometry of governance in which unilateral and bilateral initiatives outside the AC and involving the Arctic states, non-state actors and key non-Arctic states drive a more fragmented development of the region.

*Is regional multilateralism in the Arctic sustainable and are there lessons here for other regions? What steps would be needed to strengthen multilateral governance in the Arctic or are states the most effective institutions for managing the region?*

## Arctic exceptionalism and global governance

The Arctic is subject to several global frameworks that have a direct impact on the region, or significant parts of it, with UNCLOS being the primary legal framework regulating the Arctic Ocean. Despite the fact that the USA continuously fails to ratify it, all Arctic littoral states have agreed to follow UNCLOS and recommendations made based on it.

Arctic fishing and shipping are both regulated in global conventions: the Polar Code (as regards amendments for the International Convention for the Safety of Life at Sea Convention, SOLAS) was approved in the autumn of 2014; and amendments to the International Convention for the Prevention of Pollution from Ships (MARPOL) were approved in May 2015. According to the Arctic states, these conventions together with national legislation are sufficient to govern the region. An Antarctic-style treaty is forcefully dismissed. At the same time, Arctic states do accept the concept of the 'common heritage of mankind' (hereafter referred to as the 'common heritage of humanity') as established by UNCLOS. They do so by setting the outer limits of their continental shelf, which does not extend across the entire Arctic Ocean, through the UNCLOS process of delimitation. Everything that remains beyond the shelf limits will be acknowledged as common heritage and administered by the International Seabed Authority. With the

Arctic states setting the rules of the game in the region, states outside the Arctic are concerned about some of the changes taking place there as well as the perceived desire to disregard non-Arctic states' interests.

While much attention has been focused on the creation of a regional framework for Arctic governance, a key issue—and sometimes a source of friction—is how the Arctic will fit into wider global issues. This is especially important in the context of claims, largely from non-Arctic states and international NGOs, that the international community has an interest in Arctic issues.

*What is the balance between international and regional governance of the Arctic? How should the fact that the changes taking place in the Arctic have serious global impacts be taken into account? Does the Arctic constitute an 'exceptional' region in the international order due to its special characteristics, notably environmental fragility? If so, does this strengthen the case for regional governance?*

### Security and governance in the Arctic

A 2012 SIPRI study showed that there was little evidence of a new militarization of the Arctic and the Arctic states did not realistically see the prospect of serious combat in the region.[1] International exchanges and circumpolar forums on defence issues in recent years have primarily focused on soft security issues, for example SAR and countering human trafficking, that continue to largely set the security agenda today. Thus, the relevant focus is not necessarily the state but rather transnational and non-state actors, and the new security agenda of organized crime, terrorism, energy security, climate change and environmental threats.

The current Arctic security architecture consists of a patchwork of national defence structures, international alliances and security organizations, which together promote varying degrees of military integration in the region. The deeply integrated North American region, the transatlantic cooperation in the North Atlantic Treaty Organization (NATO) and the, to a lesser extent, integrated Nordic countries all have stakes in the Arctic. These diverse and often

---

[1] Wezeman, S. T., *Military Capabilities in the Arctic*, SIPRI Background Paper (SIPRI: Stockholm, Mar. 2012).

overlapping structures suggest that the Arctic is not a single security complex, but rather at the intersection of several complexes. It has been suggested that any conflict in the region is more likely to originate elsewhere and to use the Arctic as a proxy.

The growing confrontation between Russia and the transatlantic community—focused on Ukraine but spreading far wider—represents a potentially major challenge to the existing patterns of Arctic security and relations between states. The crisis in Ukraine has begun a complex and mutual process of readjusting strategic assessments and even force postures between Russia and the transatlantic community. The outcome of this shift is unpredictable but it is already spilling into the Arctic.

In the post-cold war period, political relations in the Arctic were decoupled from the continuing presence of strategic nuclear forces in the region. This move enabled the emergence of the 'spirit of cooperation', which has underpinned the rise of new cooperative governance in the Arctic. With the conflict in Ukraine prompting a rethink of European security, the strategic significance of the Arctic for the Euro-Atlantic community and Russia has again come back into focus. Sweden and Finland are re-examining territorial defence, which will inevitably involve their Arctic regions. NATO countries are once again looking at whether they need to make new preparations to operate in the Arctic.

There seems to be little appetite, at present, for a renewed military confrontation in the Arctic, but any new security investments in the region are likely to be viewed with a different perspective from recent years. And a certain 'bleeding' of suspicions into the Arctic is likely.

*Does the Arctic constitute a distinct security region or should some of it be viewed primarily as part of other security regions? How do the diverse forms of security cooperation in the Arctic influence its governance and development? Is there a need, and is there the scope, to strengthen security governance in the Arctic? What forms could such cooperation take? How will security relations in the Arctic be affected by confrontation between Arctic states elsewhere?*

## III. The structure of the report

This report addresses the key themes and questions relating to Arctic governance through a set of seven chapters written by leading scholars and experts in the region. Following this introduction, in chapter 2, 'Security in the Arctic: definitions, challenges and solutions', Alyson Bailes addresses the issue of defining security in the Arctic. She notes that the Arctic is not a discrete security space but rather is co-opted into a variety of other security spaces.

In chapter 3, 'Understanding national approaches to security in the Arctic', Kristofer Bergh and Ekaterina Klimenko chart the perceptions that inform approaches to Arctic security within the Arctic subregions: North America (Canada and the USA), Europe (the Kingdom of Denmark, Finland, Norway and Sweden) and Russia, which stands on its own in the Arctic. The diverse perspectives confirm Alyson Bailes' conclusion that there is little prospect of the emergence of a 'genuine security community' in the Arctic in the short term.

The Ukraine crisis has critically affected the core relationships of the Arctic states and introduced the prospect of security developments in the region being part of the wider confrontation between Russia and the transatlantic community. This raises the important issue of whether the cooperative spirit can continue in the Arctic. In chapter 4, 'Russia's Arctic governance policies', Andrei Zagorski outlines the key approaches to international cooperation that underpin Russia's position on Arctic questions. Given Russia's pivotal geographic location and the large number of Arctic territories under Russian jurisdiction, the country's position has been vital to the emergence of cooperation on Arctic issues.

In chapter 5, 'North East Asia eyes the Arctic', Linda Jakobson and Seong-Hyon Lee consider the rising interest in the Arctic among three North East Asian states—China, Japan and South Korea. The emergence of a set of non-Arctic states with a keen and growing interest in the Arctic has been one of most notable developments over the past decade. While there has been a strong history of scientific work in North East Asia on polar regions, the challenges and opportunities posed by climate change have been the catalyst for a wider engagement in Arctic affairs.

In chapter 6, 'The Arctic Council in Arctic governance: the significance of the Oil Spill Agreement', Svein Vigeland Rottem explores what the agreement on oil spill preparedness and response, negotiated under the auspices of the AC, indicates about the balance between international and national governance in the Arctic with respect to the core interest of energy. The emergence of the AC as a significant international institution over the last decade is, perhaps, the most visible indicator of the creation of new forms of Arctic governance. The final chapter summarizes the main arguments in each chapter and draws some conclusions from the discussion. Chapter 7 also provides an analysis of the status and future prospects of Arctic governance, which, after two decades of growing regional and international interest, constitutes a sensitive balance between international, regional, national and informal governance structures.

# 2. Security in the Arctic: definitions, challenges and solutions

ALYSON J. K. BAILES*

## I. Introduction

The Arctic is not a distinct and unitary security space, nor is it an intact or 'virginal' one. It has long been co-opted into larger patterns of military activity, including the strategic confrontation and balance between the United States and the North Atlantic Treaty Organization (NATO) on the one hand, and the Soviet Union/Russia and its partners on the other.[1] Arctic dynamics in 'softer' areas of security—environmental, economic, energy, infrastructure, societal and human security for instance—are heavily influenced by transnational phenomena and far-reaching interdependencies linking the region with the whole northern hemisphere, or even the globe. At the same time, Arctic security capabilities, activities and relationships are fragmented between different national jurisdictions, bilateral relationships and groupings. The circumpolar space is a set of subregions, with striking contrasts in formal institutional coverage between—notably—the North Atlantic gateway, and the North Pacific approaches and the Bering Sea.[2]

Such complex and confusing governance arrangements do not in themselves make a region, its states or its peoples insecure. Indeed, it is easiest to portray the Arctic as a region of potential conflict when one strand of the tapestry is singled out without considering complicating and balancing factors. The fact that Arctic security

---

* The author thanks doctoral candidate Gustav Pétursson for valuable research support.

[1] For a summary of historic military exploitation of the Arctic see le Mière, C. and Mazo, J., *Arctic Opening: Insecurity and Opportunity* (IISS Adelphi Series, Routledge: Abingdon, Dec. 2013), pp. 78–83.

[2] The North Atlantic region has multilayered structures from the NATO–Russia and EU–Russia relationships down to localized groupings like the Barents Euro-Arctic Council and the EU's Northern Dimension. Difficulties including Russia–Japan territorial disputes have obstructed anything beyond coastguard and fisheries cooperation in North East Asia. On Arctic governance in general see Schram Stokke, O. and Hønneland, G., *International Cooperation and Arctic Governance* (Routledge: London, 2009).

challenges are manifold and rapidly evolving is not untypical either. Globalization, combined with new technologies and patterns of human behaviour, makes the future of every world region contingent on successful *management* of change. This in turn demands appropriate governance solutions, with the emphasis on 'appropriate'. As with appropriate technology, what will best serve future peace, human welfare and conservation of the biosphere will not necessarily be familiar tools, nor those normally considered the 'strongest' and most advanced.

The forces of change have also transformed contemporary understandings of security. Both in academic analysis and in national and international policymaking, modern definitions of security are commonly multidimensional.[3] They still include military security and defence, but embrace many other phenomena affecting the safety and wellbeing of nations, societies and individuals: from terrorism and crime, through accidents, infrastructure failures and interruption of vital supplies, to general economic or social weaknesses, natural disasters and climate change, and pandemic disease. The Arctic strategies of concerned states and institutions (chapter 3 in this volume) cover all these topics and more. This chapter follows a similar approach, grouping the relevant security issues into four interconnected 'packages': (*a*) 'hard' security and conflict; (*b*) environmental, energy and economic security; (*c*) the handling of civil emergencies; and (*d*) aspects of societal and human security.[4]

Such broad and multiform definitions have practical as well as conceptual consequences. They mean that a much wider range of international organizations, and other interstate or non-state transactions, than previously can be seen as affecting security for good or ill. The non-military dimensions of security depend heavily on the private economy (including its supply systems, infrastructures

---

[3] Examples of academic analysis include Collins, A. (ed.), *Contemporary Security Studies* (2nd edn) (Oxford University Press: Oxford, 2010); and Williams, P. D. (ed.), *Security Studies: An Introduction* (Routledge: London, 2008). Examples of international policymaking are the European Union's Security Strategy, 'A secure Europe in a better world', Brussels, 12 Dec. 2003, <http://www.consilium.europa.eu/uedocs/cmsUpload/78367.pdf>; and NATO's new Strategic Concept, 'Active engagement, modern defence', adopted by heads of state and government at the NATO Summit in Lisbon, 19–20 Nov. 2010, <http://www.nato.int/strategic-concept/pdf/Strat_Concept_web_en.pdf>.

[4] The terms 'societal' and 'human' security are explained in section V below.

and essential services) and on the natural or man-made environment, thus greatly increasing the range of non-state actors that must similarly be recognized as having a security role. Currently, at a time when defence industries are often privatized and most armed violence takes place between combatants within states, the relative importance of non-state actors in security is rising across the board. It strengthens the case for critically reviewing traditional methods of security governance.[5]

This chapter aims to illustrate the complexity both of security agendas and the definition of security actors in the modern Arctic, while highlighting codependencies between security conditions there and elsewhere. Sections II to V identify the challenges, and explore the range of relevant actors and governance instruments in each of the four security packages mentioned above. Section VI discusses the subsequent need for multidimensional and multilevel security solutions, and draws the analysis to a conclusion.

## II. Dimensions of Arctic security: hard security and conflict

The implications of Arctic change in the field of traditional, military defence have been intensely debated in recent years but have led to divergent conclusions. A group of Canadian researchers warned in 2012 that 'many of the Arctic states' actions and statements make it clear that they intend to develop the military capacity to protect their national interests in the region ... the Arctic nations will reserve the right to use unilateral force to defend their interests if necessary'.[6] A SIPRI study from the same year, however, stated that Arctic states were making only 'limited increases in their capabilities to project military power beyond their recognized national territories', and that 'Conventional military forces specially adapted to the harsh Arctic environment ...

---

[5] See Themnér, L. and Wallensteen P., 'Patterns of organized violence, 2003–12', *SIPRI Yearbook 2014: Armaments, Disarmament and International Security* (Oxford University Press: Oxford, 2007), pp. 70–89.

[6] Huebert, R. et al., *Climate Change and International Security: The Arctic as a Bellwether* (Center for Energy and Climate Solutions: Arlington, VA, May 2012).

will remain in some cases considerably below cold war levels'.⁷ In December 2013 the International Institute of Strategic Studies concluded that 'the reality is that the Arctic is not witnessing an uncontrolled or substantially competitive militarization . . . it is far from being a battleground for rival states'.⁸

These differences do not reflect uncertainty over the facts. The key difference between the Arctic and the Antarctic is that five states—Canada, the Kingdom of Denmark, Norway, Russia and the USA—have significant land territories and maritime jurisdictions above the Arctic Circle. They are commonly called the five littoral states, although Iceland, situated just below the Arctic Circle, also claims littorality under the United Nations Convention on the Law of the Sea (UNCLOS).⁹ All six states are set on upholding their sovereignty and benefiting from ownership of all resources within their national purview. The regional system is thus one of 'modern' or 'realist' politics, rather than an Antarctic-style *terra nullius* or a post-modern community of modified sovereignty such as the European Union (EU). Accordingly, the balance of competition and cooperation is shaped primarily by intersecting national interests, even if all states proclaim more altruistic concerns (e.g. the environment or indigenous peoples).

Further, all the states of the region except Iceland possess military forces and have plans to modernize and/or expand their Arctic-related capabilities. For Russia, these include new ship and submarine construction, the new subordination of an Arctic brigade based in Pechenga to the Northern Fleet, and the reactivation of two cold war bases in the New Siberian islands and Severnaya Zemlya, with more changes likely to come.¹⁰ Since 2007, Russian strategic bomber aircraft have also patrolled more frequently and widely around Iceland and other Nordic territories. Canada has ordered 6–8 ice-strengthened offshore patrol ships. Norway has moved its national defence headquarters northwards to Bodø and

---

⁷ Wezeman, S. T., *Military Capabilities in the Arctic*, SIPRI Background Paper (SIPRI: Stockholm, Mar. 2012).
⁸ Le Mière and Mazo (note 1).
⁹ United Nations Convention on the Law of the Sea (UNCLOS), opened for signature 10 Dec. 1982 and entered into force 16 Nov. 1994, <http://www.un.org/depts/los/convention_agreements/texts/unclos/unclos_e.pdf>.
¹⁰ See Wezeman (note 7) and Le Mière and Mazo (note 1).

is procuring new ships and Joint Strike Fighter aircraft. Denmark formed a new Arctic Command at Nuuk, Greenland, in 2012 and has been strengthening its fleet of frigates and patrol vessels. Only the USA has so far held back on new Arctic procurement, while designating the head of its Northern Command as 'Arctic advocate' and seeking benefits from better governmental and public–private coordination.[11]

The difficulty lies in judging what conclusions to draw from these observations. While some commentators have tactical motives for amplifying or understating the threat of conflict, there are two genuine elements of ambiguity in the picture presented by Arctic military developments.[12] Firstly, the region's states differ, not only in their size and scope for military power play, but in their strategic outlooks and the part that Arctic deployments play within these. The Nordic states have, at best, defensive military capabilities, and no ambitions beyond their own territory except for supporting peace missions. Canada, while often drawing attention to signs of Arctic military competition, is not adopting an aggressive or expansionist posture.[13] For the USA, present only through Alaska, the Arctic is a peripheral concern given its other geographical outlets and strong strategic engagement in other regions.[14]

For Russia, by contrast, the Arctic seas provide one of its few remaining strategic 'break-out' zones and are central to its transmuted, but still tense, nuclear and naval confrontation with the USA and NATO.[15] Russia's Northern Fleet is the country's largest, and includes two-thirds of its total submarine capacity.[16] It is also

---

[11] US Department of Defense, 'Arctic Strategy', Nov. 2013, <http://www.defense.gov/Portals/1/Documents/pubs/2013_Arctic_Strategy.pdf>.

[12] Leaders who 'talk tough' on the Arctic may be trying to deter other states from encroaching on their interests, but also to justify new military spending and/or win political support, notably from northern constituencies.

[13] Bergh, K., 'The Arctic policies of Canada and the United States: domestic motives and international context', SIPRI Insights on Peace and Security no. 2012/1, July 2012, <http://books.sipri.org/product_info?c_product_id=446>.

[14] Wezeman (note 7).

[15] The Northern Front's salience grew in 1989–90 when Russia lost its European allies and troop positions with the Warsaw Treaty Organization's dissolution. In the Baltic and Mediterranean, surrounded by NATO or NATO-friendly states, Russia now has less room for major anti-Western manoeuvres.

[16] Le Mière and Mazo (note 1), p. 84.

here that Russia sees the most immediate challenge to strategic balance from the USA's continuing, albeit watered-down, ballistic missile defence programme.[17] Thus, for the Russians, even if activity levels slumped after the cold war, the Arctic never lost its crucial role in balance and mutual deterrence with other great powers, including Asian neighbours. Today, when Russia is pouring resources into defence, having more than doubled its military expenditure since 2003, it would be remarkable if no further improvements were scheduled on the northern front.[18] The important point to grasp is that Russian actions and gestures made in the northern theatre may have little or nothing to do with the Arctic as such. When they reflect strategic rivalry with the USA, NATO, or other world powers, any response directed at Russia may more appropriately and effectively be made outside of the Arctic. Overall, and not least in the light of recent events in Ukraine, it may be just as (or more) realistic to imagine conflict spreading into the Arctic from elsewhere than to think of it starting there.

The second ambiguity lies in the multiple roles played today by armed forces and their specialized assets. Particularly relevant in the Arctic are the functions of sovereignty support—demonstrating a state's ability to patrol, monitor and act across its whole land and sea territory—and support to the civil power for purposes such as shipping escort and fisheries protection, combating crime and smuggling, search and rescue (SAR) operations, and assistance in civil emergencies such as large accidents and natural disasters. Both kinds of service are more in demand throughout the Arctic today as the ice-free, accessible parts of national jurisdictions expand, and human activity grows with all that it implies for new infrastructures, extended communications, and accident and pollution risks (discussed further in section IV below). They may be performed by traditional armed forces as well as paramilitary or constabulary organizations, such as coastguards or the Canadian Mounted Police. How far can planned military increases be

---

[17] Dvorkin, V. Z., 'Missile defence and security in the Arctic', ed. A. V. Zagorsky, *The Arctic: A Space of Cooperation and Common Security* (IMEMO Ran: Moscow, 2010).

[18] SIPRI Military Expenditure Database, <http://www.sipri.org/research/armaments/milex/milex_database/milex_database>. In fact, other Russian fleets are scheduled to benefit more than the northern one from new naval construction. Le Mière and Mazo (note 1), p. 85.

SECURITY IN THE ARCTIC 19

explained by such needs, which involve a normal exercise of sovereignty and need not be seen as aggressive or destabilizing? It is hard to assign a given unit or asset firmly to one role or the other. Tasking is often kept loose, both for the sake of flexibility and deterrence, especially in the case of highly mobile naval and air forces, which account for most of those active in the Arctic. The SIPRI study cited above, however, plausibly concludes that such 'civil' uses of armed forces account for much planned Arctic military development, while some acquisitions merely replace obsolete items.[19]

Most current armed conflicts have their origins in violence between groups within the state, whether driven by political, ideological, socio-economic or separatist motives.[20] Like terrorism, they flourish in 'weak state' conditions where the state's monopoly of force is compromised and borders are poorly defended against smuggling and infiltration. Internal disorder invites meddling by neighbours (for instance, Ethiopia in Somalia) and other interested parties, and has been linked with state-on-state attacks in such prominent recent cases as Afghanistan. It is noteworthy, therefore, that the whole Arctic region is remarkably free from all such phenomena. The states composing it are 'strong' rather than 'weak' and are rich by world standards. Local populations are small; and while divisive issues and grievances exist, notably over indigenous rights, there is no modern tradition of intra-societal violence or terrorism in any Arctic territory. No Arctic state has shown any inclination to take up the cause of an anti-government group in an Arctic neighbour's northern provinces, let alone to intervene with armed force.[21]

Among the national strategies of Arctic Council (AC) members, the US strategy from 2009 devotes particular attention to potential dangers posed by violent non-state actors, and to the risk of the Arctic being exploited by international terrorists or pirates.[22] A

---

[19] Wezeman (note 7).
[20] Themnér and Wallensteen (note 5).
[21] Bailes, A. J. K., 'Human rights and security: wider applications in a warmer Arctic?', *The Yearbook of Polar Law*, vol. 3 (Martinus Nijhoff: Leiden, 2011).
[22] White House, National Security Presidential Directive and Homeland Security Presidential Directive, Arctic Region Policy, 9 Jan. 2009, <http://georgewbush-whitehouse.archives.gov/news/releases/2009/01/20090112-3.html>.

more immediate scenario is that of sabotage or other forceful action by non-governmental campaigners such as Greenpeace, who attempted to board a Gazprom oil-drilling rig in the Pechora Sea in 2013.[23] Could such incidents trigger international conflict? On the contrary, Arctic states and local communities should have a common interest in disciplining actors that knowingly defy local laws. The Greenpeace incident itself illustrated governments' reluctance to risk conflict and retaliation by questioning each other's jurisdiction in such cases. Non-governmental campaigns using more peaceful methods may succeed better in agenda setting and mobilizing political forces, but in the 'strong state' environment of the Arctic, handling them should be a political and legal challenge rather than cause for violence.

Turning to governance, the Arctic ostensibly lacks institutions with hard security competence. The AC's members agreed at its inception that defence issues would be absent from its agenda, as they are in the more localized groupings created in the 1990s to foster West–Russian cooperation in the Arctic: the Barents Euro-Arctic Council (BEAC) and the EU's Northern Dimension.[24] The Arctic states do not share an alliance or other traditions of region-wide military cooperation. Nor have they created a regional arms control regime or a set of confidence- and security-building measures (CSBMs) designed for and covering the whole Arctic space. True, all are members of the Organization for Security and Cooperation in Europe (OSCE); but the conventional arms control agreements negotiated within that institution's framework were designed to cover the European continent 'from the Atlantic to the Urals' and have never placed restrictions on naval forces. OSCE CSBMs, which provide notably for the notification and observation of military exercises, do apply to the 'adjoining sea area and air space' of these measures' zone of application in Europe, but only if

---

[23] For Greenpeace's version of the incident see <http://www.greenpeace.org/international/en/news/features/From-peaceful-action-to-dramatic-seizure-a-timeline-of-events-since-the-Arctic-Sunrise-took-action-September-18-CET/>.

[24] The absence of defence from its agenda is recorded in a footnote to the Arctic Council's inaugural declaration of 19 Sep. 1996, available at <http://www.arctic-council.org/index.php/en/document-archive/category/4-founding-documents>. The Barents Euro-Arctic Council (BEAC) was established in 1993, see <http://www.beac.st/>. The Northern Dimension is a joint policy between the EU, Norway, Russia and Iceland, which was initiated in 1999, see <http://www.eeas.europa.eu/north_dim/>.

the activities involved are linked to European security.[25] The Open Skies Treaty allows all OSCE participating states to mount observation flights over each other's territory—including Canada and Alaska—but only as far out to sea as these states' territorial waters extend (a maximum of 12 nautical miles).[26]

To conclude that Arctic defence relationships know no law or restraint is, however, an unwarranted leap of logic. Four littoral states, and Iceland, are members of NATO, and Russia has a formal relationship with NATO designed to provide lines of communication even during more confrontational episodes. Insofar as Russia's military efforts in the Arctic serve larger considerations of East–West rivalry and balance, they must logically also be constrained by the nuclear-based deterrence that has arguably guaranteed peace with the West since the 1940s. Russia has no reason to think it could attack Norwegian territory, Icelandic sea space or Canadian vessels, any more than it could invade Estonia, with impunity. This inhibition might theoretically be weaker in circumpolar sea areas where littoral states' current claims overlap, but—aside from the fact that the claimants have agreed to act peacefully and legally—there are no predicted seabed or fish resources there that would be worth taking the risk for.[27] Accidental clashes in the Arctic could be addressed through general-purpose 'hot-line' arrangements. Military scenarios not involving Russia are difficult to imagine: other jurisdictional disputes involve fellow members of NATO who are most unlikely to attack each other in such a cause. No non-Arctic power has apparent plans or means to inject a military presence; and China, often cited as a possible disruptive force, is also part of the system of mutual nuclear deterrence among the great powers.[28]

---

[25] 'Sea area' was later defined also to refer, where necessary, to any relevant 'ocean area'. The zone of application is most fully explained in Annex I of the Vienna Document 1999, <http://www.auswaertiges-amt.de/cae/servlet/contentblob/388296/publicationFile/4108/WienerDok-E.pdf>.

[26] The Treaty on Open Skies entered into force on 1 Jan. 2002, <http://www.osce.org/library/14127>.

[27] The 5 littoral states did so when meeting at Ilulissat, Greenland, on 27–29 May 2008. The declaration text is available at <http://www.oceanlaw.org/downloads/arctic/Ilulissat_Declaration.pdf>. On environment, energy and economics, see section III below.

[28] These arguments are developed further in Bailes, A. J. K., 'Turning European security upside down? The future significance of the Arctic', Dis-politika, vol. 37/3–4, Feb. 2013. On China see Jakobson, L., 'China prepares for an ice-free Arctic', SIPRI Insights on

NATO's role in the Arctic is best understood in this perspective. Throughout the cold war, Norway and Denmark made sure that NATO avoided unnecessary provocation by eschewing exercises near the Soviet border and declining to station foreign troops or nuclear items on Soviet territory. When NATO held a high-level Arctic seminar in Reykjavik in 2009, the chair concluded that any development of its activities should be limited to monitoring, situational awareness and help in civil emergencies.[29] Up to 2015, Canada was in any case—for its own reasons—blocking the development of an Arctic strategy or activity programme in NATO (although coming under increasing pressure to reconsider this stance).[30] Russia, for its part, has a strong and vocal preference for keeping things this way, just as it opposes NATO installing new bases close to its Western land borders. None of this, however, alters the fact that NATO's guarantees extend to the Northern limit of its members' jurisdiction.[31] By restricting itself (thus far) to a largely 'over-the-horizon' presence plus occasional exercises, NATO has not foregone its stabilizing role of deterrence. Rather, at least up to 2014, its members were content to leave room in the Arctic both for positive non-military interactions, and for other, less contentious forms of military cooperation.

Russia has carried out joint naval exercises with Norway and the USA, and takes part in pan-Arctic consultations on military support for civil emergencies—covered below. The USA and Canada have long shared an air defence system (NORAD) and are developing Arctic cooperation in 'planning, domain awareness, information-sharing, training and exercises, operations, capability development and science and technology'.[32] Since US forces left its territory in 2006, Iceland has made bilateral defence agreements

---

Peace and Security no. 2010/2, Mar. 2010, <http://books.sipri.org/files/insight/SIPRI Insight1002.pdf>.

[29] Chairman's conclusions from the NATO conference 'Security prospects in the High North', Reykjavik, 28–29 Jan. 2009.

[30] Canada typically argued that Arctic affairs should be left to the littoral and local states to manage. This line became hard to hold in light of Ukraine-related tensions with Russia, including the tangible impact of Western sanctions on planned West–Russia activities in the Arctic, and was expected to be reconsidered by a new Canadian government elected in October 2015.

[31] NATO's operational area as defined in the 1949 Washington Treaty has no northern limits.

[32] Canadian Government, 'The Canada–U.S. defence relationship', Backgrounder, Nov. 2013, <http://www.forces.gc.ca/en/news/article.page?doc=the-canada-u-s-defence-relationship/hob7hd8s>.

with Denmark, Norway, Canada and the United Kingdom, and hosts periodic air patrolling/policing exercises by NATO, the latest of which (February 2014) was joined by Finnish and Swedish aircraft.[33] Defence cooperation among the five Nordic states, including in regional operations, has steadily increased in recent years, as signalled by the Nordic Defence Cooperation (NORDEFCO) agreement of November 2009.[34] The Stoltenberg Report presented that year included several proposals for directing such cooperation towards the Arctic region, one of which was fulfilled by the aforementioned Finnish–Swedish participation in the air policing exercise.[35]

Such positive cooperation has its own benefits for regional stabilization, understanding and transparency. Could it be seen as fulfilling some of the functions of traditional arms control and confidence building, given the problems of applying the latter to naval and air forces? While interest in Arctic arms control ebbed after the cold war, the case for creating specific regional CSBMs is still supported by some analysts, including the SIPRI report cited earlier.[36] It might be considered more pressing, as it is for the whole Euro-Atlantic space, in the light of recent Russian manoeuvres against Ukraine; not to mention the longer-term advantages of extending rules of transparency and restraint to other (notably Asian) powers looking to engage in Arctic waters.[37]

## III. Environment, energy and economics

The concepts of environmental security and economic security might appear to be conflicting, perhaps even mutually exclusive, especially as defined by their most vocal adherents. Yet the natural world and the human economy are intrinsically linked, since the latter cannot function—even at the most primitive level—without the former's resources, and environmental protection draws in

---

[33] NATO Allied Command Operations, 'NATO and partner fighter jets soar over Iceland', 14 Feb. 2014, <http://www.aco.nato.int/nato-and-partner-fighter-jets-soar-over-iceland.aspx>.
[34] For further details on NORDEFCO see <http://www.nordefco.org/>.
[35] Norwegian Government, 'Stoltenberg Report presented to Nordic foreign ministers', 9 Feb. 2009, <www.regjeringen.no/en/dep/ud/Whats-new/News/2009/nordic_report.html?id=545258>.
[36] Wezeman (note 7).
[37] For further details see chapter 5 in this volume.

turn on human-created wealth. In the Arctic, climate change is opening up new spaces for potential economic exploitation, including fossil fuel extraction, minerals mining, freight shipping, fisheries and tourism, together with the related logistical and support services. The difficulty is that any such activities must draw directly or indirectly on the Arctic's natural resources, with the risk of misusing, spoiling or prematurely exhausting them. Even the expansion of non-extractive sectors such as shipping aggravates risks of pollution that, in a negative feedback loop, could further accelerate global warming. Beyond such practical concerns, the ecological lobby also sees it as normatively wrong to raid the resources of one of the world's last great wildernesses, already ravaged by extraneous pollution and the man-made drivers of climate change.

The environment and the economy are not security issues as such, but their prominence in Arctic states' policy documents shows that governments see some potential implications with strategic or even life-and-death significance. Mishandling of environmental, energy and other economic processes can put human lives as well as livelihoods in jeopardy, and threaten the viability of the state itself. There is also a putative link between these processes and the causation of both intrastate and interstate conflict. Both aspects are considered here.

### Environmental security

The concept of environmental security has evolved since the 1970s from a focus on protecting the environment itself, to a more complex acknowledgement of the human dimension. People benefit from nature, but need to be protected from its violence and extremes.[38] Climate change is raising the stakes in the Arctic in all these respects. The melting of sea ice is destroying wildlife habitats and the hunting grounds of indigenous peoples, while rising sea levels threaten coastal settlements. On land, the melting of permafrost layers destabilizes any structures built on them, a major

---

[38] Ingólfsdóttir, A. H., 'Environmental security and small states', eds. C. Archer, A. J. K. Bailes and A. Wivel, *Small States and Interational Security: Europe and Beyond* (Routledge: London, 2014).

challenge for Russia's Siberian industrial zones and lines of communication. Weather patterns and sea currents are changing, creating more extreme temperatures and wind speeds and new distributions of icebergs, as well as a general increase in unpredictability and the risk of natural disasters.[39]

These processes are already impeding human activities in the Arctic, imposing burdens and losses that are difficult for the less developed areas to absorb. Plans for new commercial activity, especially along the coasts of Alaska, Canada, Greenland and Siberia, are driven partly by the need to find compensating sources of wealth and new employment prospects for communities bereft of traditional lifestyles.

## Economic security

Such proactive adaptation fits the logic of economic security, which requires the assured creation or supply of the goods needed to keep a state viable and ensure its people's welfare.[40] Costs of mitigating climate change must also be met from somewhere. Rapid economic development, however, risks boosting pollution, hastening resource exhaustion and raising levels of climate-sensitive emissions. Economic and financial security can itself be prejudiced if Arctic schemes become ill-considered 'bubbles' without reliable funding and insurance, or if 'hot money' is drawn into the region only to be withdrawn without concern for local interests.[41] The need for funders with deep pockets and the ability to absorb risk helps to explain the growing interest of China and other rising economies in the Arctic, and the openness to this in some local constituencies. However, when remote communities engage for

[39] Arctic Climate Impact Assessment (ACIA), *Impacts of a Warming Arctic: Arctic Climate Impact Assessment, 2004* (Cambridge University Press: Cambridge, 2004); and Intergovernmental Panel on Climate Change (IPCC), *Climate Change 2014: Impacts, Adaptation and Vulnerability*, Summary for Policymakers (Cambridge University Press: Cambridge and New York, NY, 2014).

[40] Buzan, B., 'Economic security', *Peoples, States and Fear: An Agenda for International Security Studies in the Post-Cold War Era* (2nd edn) (Lynne Rienner Publishing: Boulder, CO, 1991), pp. 230–61.

[41] Iceland's financial crash in 2008 exemplified both a collapsing 'bubble' (of unsound investment and debt) and the flight of 'hot money'. If incoming investment fails to promote local skills and employment, places like Greenland could become 'rentier' societies with their own problems of stability and morale.

the first time with much larger countries and/or powerful multinational companies, there are obvious risks of weak regulation and enforcement, undue political influence and aggravation of the societal problems discussed in the subsection on societal and human security below.[42]

### Energy security

Energy security preoccupies both net importers and suppliers of fuels and electricity. The former seek assured supplies at manageable prices, preferably without political or strategic strings attached, while the latter must balance short-term profit seeking with their need for security of demand from satisfied partners.[43] Among Arctic states, Norway and Russia are established oil and gas exporters, while Iceland is 80 per cent energy self-sufficient thanks to hydropower and geothermal resources. Exploiting new Arctic resources offers the Norwegians and Russians insurance against the exhaustion of existing stocks, and might give Russia new, logistically easier options (though with heavy up-front costs) for exporting to China and Japan. These two countries duly define the energy dimension as a top Arctic priority. For Greenland and the Faroe Islands, success in developing offshore oil and gas fields could be a game changer, reducing their reliance on Danish subsidies and making full independence economically (if not politically) viable.[44] For Canada and the USA, Arctic hydrocarbon extraction should improve energy self-sufficiency. Most recently, however, the rapid rise in Canadian and US gas production through fracking processes, and wider global interest in shale oil and shale gas, have contributed to depressing both the world oil price and the strategic and commercial appeal of Arctic offshore

---

[42] Einarsson, S. K., 'China's foreign direct investment in the 'West'. Is there a security threat?', Masters thesis, University of Iceland, Reykjavik, June 2013, <http://skemman.is/item/view/1946/14799;jsessionid=E82891D4FB4C498574529E0E8CB5C422>.
[43] Proninska, K., 'Energy and security: regional and global dimensions', *SIPRI Yearbook 2007: Armaments, Disarmament and International Security* (Oxford University Press: Oxford, 2007).
[44] Nielsson, E. T., 'The West Nordic Council in the Global Arctic', Centre for Arctic Policy Studies Occasional Paper, Institute of International Affairs, University of Iceland, Reykjavik, Mar. 2014, <http://ams.hi.is/wp-content/uploads/2014/03/the_west_nordic_council.pdf>.

production, with its daunting safety and pollution risks. All evidence, including the Norwegian Government's recent halt to seabed exploration in some Arctic areas, suggests that the world energy market does not currently privilege early exploitation of Arctic energy resources.[45] In a longer-term perspective, however, the latter may still offer extra capacity for rising economies' growing needs; insure against future decline in Norwegian and Russian supplies; and help to ease Western and Asian reliance on sources in the Arab world.

Any such benefits depend on Arctic energy stocks not igniting conflict, or being used in aggressive power games. Three main factors militate against such scenarios. The first, already noted, is the general inhibition on military force among countries within the NATO–Russia deterrence regime. Secondly, and as briefly noted in section II, the five littoral states declared in May 2008 that they would: (*a*) respect the 'legal framework' for Arctic development; (*b*) seek 'orderly settlement' of outstanding claims; and (*c*) ensure 'responsible management' of the region and its resources.[46] The third factor is the distribution of the predicted deposits. Maps from the US Geographical Survey (USGS) show them lying almost exclusively in land and sea areas under the uncontested jurisdiction of littoral states.[47] Current exploration is based on licences issued by the relevant governments, and consortiums including foreign companies have gained a good share of these—notably, the US firm ExxonMobil in a number of new fields off Siberia.[48] This seems to signal mutual readiness to share risk and potential profit,

---

[45] 'Norway's new government drops Lofoten oil', Barents Observer, 1 Oct. 2013, <http://barentsobserver.com/en/politics/2013/10/norways-new-government-drops-lofoten-oil-01-10>; and Claes, D. H., 'Arctic energy resources—curse or blessing?', Arctic Governance Project Working Paper, 2010, <http://www.arcticgovernance.org/arctic-energy-resources-curse-or-blessing.4777861-142902.html>.

[46] 'The Ilulissat Declaration', Arctic Ocean Conference, Ilulissat, Greenland, 27–29 May 2008, <http://www.oceanlaw.org/downloads/arctic/Ilulissat_Declaration.pdf>.

[47] Constantly updated materials are available on the USGS webpage, 'Circum-Arctic Resource Appraisal', <http://energy.usgs.gov/RegionalStudies/Arctic.aspx>.

[48] ExxonMobil, 'Rosneft and ExxonMobil advance strategic cooperation', News and updates, 21 June 2013, <http://news.exxonmobil.com/press-release/rosneft-and-exxonmobil-advance-strategic-cooperation>. At the time of writing, however, the execution of these contracts was being called into doubt by Ukraine-related US sanctions that were seeking to halt the transfer of Western expertise to Russian oil and gas industries, including those in the Arctic.

rather than the start of a mercantilist energy war. Few, if any, valuable resources have been predicted in the circumpolar area where nations' outstanding claims overlap, and where extraction conditions will be especially uncertain and difficult for some time yet. In the market mood described above, economic motives would seem among the weakest for risking a show of force there.

In governance terms, these issues provide a prime example of challenges affecting the Arctic that cannot be mastered by the Arctic's own inhabitants and institutions. The anthropogenic factors driving global warming lie overwhelmingly outside the Arctic Circle. Mitigating them is a challenge for the major emitters in North America, the EU, Russia, China and Japan, and for the series of world climate talks held under UN auspices.[49] The role of the AC has been one rather of diagnosis, publicity and advocacy: from the Arctic Environmental Protection Strategy that its members-to-be endorsed in 1991, to its present research on damage from black carbon (soot) deposits.[50] Adaptation to climate change is primarily a national responsibility, and the Arctic is unlikely ever to benefit from international climate-related aid programmes. However, the AC is active in exploring impacts and options notably in its human development work (see below), and the BEAC and the EU's Northern Dimension can fund cooperation on solutions ranging from more resilient infrastructures to tourism management. There is a separate tradition of Western–Russian–Japanese cooperation to deal with nuclear pollution, mostly military-generated and a severe problem in some Arctic locations.[51] At the non-governmental level, links between the indigenous peoples' organizations provide for bottom-up debate and exchange of ideas.

In globalized free-market conditions, the main determinants of Arctic economic development lie outside governmental and institutional control. As the UK's recently published Arctic policy

---

[49] For further details see the UN Framework Convention on Climate Change website, <https://unfccc.int/>.
[50] The strategy text is available at <http://www.arctic-council.org/index.php/en/document-archive/category/4-founding-documents>. Arctic Council, 'The Task Force for Action on Black Carbon and Methane', 27 Sep. 2013 <http://www.arctic-council.org/index.php/en/resources/news-and-press/news-archive/782-the-task-force-for-action-on-black-carbon-and-methane>.
[51] For details see eds C. Southcott and L. Heininen, *Globalization and the Circumpolar North* (University of Alaska: Fairbansk, 2010), pp 231–38.

puts it: 'Decisions on whether to proceed with exploration and extraction projects are commercial matters for operators to make in the light of prevailing market and regulatory conditions. In turn these will be affected by the prevailing environmental conditions of the area in question.'[52]

As noted above for the hydrocarbon sector, the business mood is currently cautious, due to the lingering effects of the 2008 crash and global market shifts as well as local climatic and political uncertainties, commercial and reputational risks, and high insurance costs. This offers governments a certain period of grace to apply the main tool of influence remaining to them, namely regulation. All Arctic states support the goal of sustainable development with due regard both for the environment and local rights, but all reject a comprehensive new regulatory framework modelled on the Antarctic Treaty. This leaves four main options: (*a*) national legislation, which could for instance set conditions for foreign investments and define local peoples' status in decision making; (*b*) common EU regulations and standards applied across the jurisdictions of the EU's local members and European Economic Area (EEA) members; (*c*) new sectorial regulation at global or regional level; and (*d*) self-regulation by the private sector.[53] The security potential of the last two methods is further discussed here.

Arctic shipping is a sector in long-term growth, albeit from a very low base. Freight shipping through the Northern Sea Route above Siberia has been heralded as having huge commercial potential but is handicapped by Russia's insistence on icebreaker escorts, high insurance prices and difficulty in identifying profitable cargoes to carry both ways.[54] More tangible is the growth in oil-, gas- and

---

[52] British Government, 'Adapting to change: UK policy towards the Arctic', Oct. 2013, <https://www.gov.uk/government/uploads/system/uploads/attachment_data/file/251216/Adapting_To_Change_UK_policy_towards_the_Arctic.pdf>, p. 21.

[53] On (*b*) see Koivurova, T. et al., *The Present and Future Competence of the European Union in the Arctic*, European Parliament background report, CJO 2011, doi:10.1017/S0032247411000295. Relevant EU members are Denmark (mainland only), Finland and Sweden; Norway and Iceland are EEA members.

[54] Conditions in Canada's Northwest Passage remain much more difficult. Lasserre, F., 'High North shipping: myths and realities', eds S. G. Holtsmark and B. A. Smith-Windsor, *Security Prospects in the High North: Geostrategic Thaw or Freeze* (NATO Defence College: Rome, 2009).

mining-related traffic from fields already being developed and in tourist cruises. With regard to the latter, the number of passengers carried to Arctic destinations in 2004 (1.2 million) had more than doubled by 2007.[55] Further expansion offers both benefits and risks. The benefits would accrue to locals providing support and entrepôt services as well as to the shippers, but are offset by environmental security risks arising from accidents that may release pollutants, as well as spillages and emissions in daily operation. The MARPOL convention of the International Maritime Organization (IMO) deals globally with pollution prevention, and coastal states can take powers under UNCLOS to designate safe shipping lanes and Particularly Sensitive Sea Areas (PSSAs).[56] However, based on a Marine Shipping Assessment Report, AC member states agreed in 2009 that a specific Arctic code for commercial shipping was needed and asked the IMO to develop it.[57] Drafting of the code, with provisions to be applied through national jurisdictions, began in 2010 and the text was finally adopted in November 2014, giving a prospect of entry into force by 2017.[58] Its provisions cover safety elements in both the design and operation of vessels. They have been equally contested by environmentalists, who feel broader ecological and societal risks have been neglected, and by operators, who fear over-strict standards will destroy their already fragile profit margins. Lloyd's of London has meanwhile proposed a voluntary code among insurers to complement the IMO provisions.[59]

In fisheries, the Arctic accounts for some 12 per cent of the world's commercial catch. The sector is of existential importance for many local communities and for nations such as Iceland and Greenland. Sea warming is expected to increase stocks overall, but migration will disadvantage some subregions and has already pro-

---

[55] Arctic Council, 'Arctic marine shipping assessment report', Apr. 2009, <http://www.pame.is/index.php/projects/arctic-marine-shipping/amsa/amsa-2009-report>, p. 79.

[56] MARPOL applies to 99% of the world's merchant tonnage. IMO, 'Pollution prevention', <http://www.imo.org/OurWork/Environment/PollutionPrevention/Pages/Default.aspx>; and Tedsen, E., Cavalieri, S. and Kraemer, R. A. (eds), *Arctic Marine Governance: Opportunities for Translatlantic Cooperation* (Springer-Verlag: Berlin, 2014).

[57] Arctic Council (note 55).

[58] For details see <http://www.imo.org/en/MediaCentre/HotTopics/polar/Pages/default.aspx>.

[59] 'Lloyd's develops Arctic ice regime to compliment Polar Code', Lloyd's, 14 Mar. 2014, <http://www.lloyds.com/news-and-insight/news-and-features/emerging-risk/emerging-risk-2014/a-common-ice-regime-for-arctic-shippers>.

voked one dispute between Iceland and its neighbours over mackerel quotas.⁶⁰ Even if such differences do not produce 'cod war' style violence, fishing grounds need protection against unlicensed intruders and stocks need protection against overfishing. Existing multilateral fishery management regimes do not cover most of the Arctic, although the North East Atlantic Fisheries Commission (NEAFC) extends to the far north in the Atlantic gateway.⁶¹ Several observers, including the EU, have advocated a moratorium on catches in newly ice-free seas.⁶² The five littoral states for their part agreed in July 2015 to abstain from commercial fishing in the jurisdiction-free 'high seas' around the North Pole, at least until full scientific assessment is possible.⁶³ Wider international discussion, in which the AC as well as competent UN organs might play a role, is to be expected on this policy approach.

Regulatory moves similar to those on shipping and fish are absent in the oil, gas and mining sectors where national legislation is paramount. The AC has played a role here by charting pollution risks and identifying best practice, for instance regarding Environmental Impact Assessments (EIAs).⁶⁴ Partly in response to public criticism, however, private corporations in the extractive sectors have developed voluntary codes addressing both environmental implications and societal–political issues, such as local consultation.⁶⁵ Following a Canadian initiative, in March 2014 the AC members agreed to establish an Arctic Economic Council where international, local and indigenous-owned businesses can consult

---

⁶⁰ BBC News, 'Mackerel quotas agreed after dispute', 12 Mar. 2014, <http://www.bbc.co.uk/news/uk-scotland-north-east-orkney-shetland-26554619>.

⁶¹ For further details on the scope of the NEAFC see <http://www.fao.org/fishery/rfb/neafc/en>.

⁶² European Union, 'Council conclusions on Arctic issues', 8 Dec. 2009, <http://www.consilium.europa.eu/uedocs/cms_Data/docs/pressdata/EN/foraff/111814.pdf>, p. 3.

⁶³ '5 nations sign declaration to protect Arctic "donut hole" from unregulated fishing', Arctic Newswire, 16 July 2015, <http://www.adn.com/article/20150716/5-nations-sign-declaration-protect-arctic-donut-hole-unregulated-fishing>.

⁶⁴ Hossain, K., Koivurova, T. and Zojer, G., 'Understanding risks associated with offshore hydrocarbon development', eds Tedsen, Cavalieri and Kraemer (note 56).

⁶⁵ Canadian Business for Social Responsibility, 'CSR frameworks review for the extractive industry', <http://www.csr360gpn.org/uploads/files/resources/CSR_Frameworks_Review_April_2.pdf>.

on all aspects of sustainable development.⁶⁶ The current Arctic situation of relatively weak commercial incentives combined with high risk awareness—fed by the explosion of BP's Deep Horizon oil rig in the Gulf of Mexico and the grounding of a Shell oil barge off Alaska in 2012–13—is propitious for further commercial self-regulation, aimed not least at keeping minor competitors out.⁶⁷ There are arguments for not limiting it to the AC circle, but drawing on existing models developed in the UN or other global contexts, so that incoming investors from Asia are equally engaged.⁶⁸

## IV. Civil emergencies

Accident risks in the Arctic, threatening lives, property and the environment, are already growing because of unstable weather conditions and more frequent extreme events. Any advance in human activity will multiply the odds on man-made and natural emergencies affecting ships, aircraft, oil/gas rigs and other industrial structures (also on land), pipelines, cables and other communication systems, as well as onshore settlements. The relevance of military assets to handling such events has already been noted. The basic problem in the Arctic is that the scale of the largest potential disasters—for instance, a cruise ship sinking with two thousand passengers—already outstrips the capacities available for response along these minimally populated coastlines. The gap between sovereign responsibility for civil protection and the ability to fulfil it is especially wide in Iceland, Greenland and the Faroe Islands, which lie in the Atlantic approaches likely to witness most of the new Arctic-related transits.⁶⁹ While local populations

---

⁶⁶ Arctic Council, 'Agreement on the Arctic Economic Council', 27 Mar. 2014, <http://www.arctic-council.org/index.php/en/resources/news-and-press/news-archive/858-agreement-on-the-arctic-economic-council>.

⁶⁷ For more on risk analysis and recommendations to producers see: Arctic Council, Protection of the Arctic Marine Environment (PAME), *Arctic Offshore Oil and Gas Guidelines. Systems Safety Management and Safety Culture. Avoiding Major Disasters in Arctic Offshore Oil and Gas Operations* (Arctic Council: Yellowknife, Mar. 2014).

⁶⁸ United Nations High Commissioner for Human Rights, 'Business and human rights', <http://www.ohchr.org/EN/Issues/Business/Pages/BusinessIndex.aspx>.

⁶⁹ On Iceland see Kaldal, S., *Assessing Emergency Capacity—Emergencies in Iceland's Search and Rescue Region* (Lund University: Lund, 2010). Faroese problems are discussed in Faroese Government, *The Faroe Islands—A Nation in the Arctic* (Faroese Prime Minister's Office: Tórshavn, Aug. 2013).

may manage to absorb the effects of natural disasters on themselves, it is not reasonable to expect them to take ownership of the safety needs of large commercial incomers. An alternative is to require companies to develop their own preventive and reaction capacity, for instance by sending tourist vessels out in pairs and providing their own rescue vessels and helicopters. This could, however, be the final disincentive to their investing in the Arctic, and any large-scale development of that kind would also stretch the capacity of local infrastructures and support facilities. Here lies the dilemma: massive expenditure up front is needed for safe and efficient development of a new Arctic economy that will only bring large profits later, if at all.

The AC cannot tackle that economic quandary, but its member states have made two legally binding agreements to address the challenges of Arctic SAR and handling major oil spills at sea, respectively. The first was signed at Nuuk in May 2011 and the second at Kiruna in May 2013.[70] Each starts by defining—without prejudice to outstanding claims—the spaces in which each Arctic state has jurisdiction and primary responsibility. They go on to discuss, and regulate practical conditions for, ad hoc assistance across national lines, not excluding military assets if needed. They prescribe ongoing consultation, extensive exchange of information and plans, and joint training and exercises. Chiefs of defence or the equivalent from AC member states met at Goose Bay in April 2012 to discuss support to the civil power in the relevant fields. They met again at Ilulissat, Greenland, in June 2013 and agreed among other things that all nations should adhere to the Marine Safety and Security Information System (MSSIS).[71] A wider Arctic Security Forces Roundtable including coastguards and UK and German representatives has been created to discuss Arctic challenges, while specialized coastguard cooperation groups exist both in the North Atlantic and North Pacific.[72]

---

[70] The texts of these agreements are available at <http://www.arctic-council.org/index.php/en/document-archive/category/20-main-documents-from-nuuk> and <http://www.state.gov/r/pa/prs/ps/2013/05/209406.htm>.

[71] 'Arctic nations set cooperation guidelines', Defense News, 27 June 2013, <http://www.defensenews.com/article/20130627/DEFREG01/306270013/Arctic-Nations-Set-Cooperation-Guidelines>.

[72] US Department of Defense, 'Eucom promotes cooperation among Arctic partners', 14 Nov. 2013, <http://www.defense.gov/news/newsarticle.aspx?id=121126>. As a result of

This Arctic-wide cooperation supplements existing efforts such as the SAR exercises held in Northern waters by NATO, some of which have been conducted in cooperation with Russia, and the increasingly close Nordic coordination on non-military security tasks known as the Haga process.[73] The Nordic countries are currently preparing a joint analysis of Arctic risks as well as a general audit of the efficiency of their cooperation. On 5 April 2011 in Helsinki, their foreign ministers made a political declaration of solidarity, obliging them to aid each other in non-military emergencies, including cyberattacks.[74] The incentives to deepen such cooperation, both regionally and Arctic-wide, are strong and likely to drive continued progress, despite problems of cost and bureaucratic obstacles. As the process directly engages military forces, it could contribute to strategic relaxation, keeping Arctic relationships focused on common interests rather than arms races or conflict risks.

## V. Societal and human security

The societal and human approaches to security have been developed since the late twentieth century as alternatives to state-centric concepts. Theorists of societal security emphasize society's collective security values, traditions, and capacities, and the importance of feelings of identity and solidarity.[75] Applied as a national policy—as currently in Norway and Sweden—societal security concentrates more narrowly on the prevention, mitigation and management of extreme events, like the civil emergencies just discussed.[76] Human security focuses on the plight of the individual and, in a detailed definition provided by the United Nations

---

Western sanctions against Russia in view of the Ukraine crisis, the Group of Eight (G8) chiefs of defence is currently suspended while the Annual Arctic Security Forces Roundtable (ASFR) has been meeting without Russian participation, most recently in Iceland in Mar. 2015. See also chapter 3 in this volume.

[73] For an example of a joint excercise see <http://www.nato.int/cps/en/natolive/news_24865.htm?selectedLocale=en&mode=pressrelease>. For further details on the Haga Declaration see <http://www.regeringen.se/sb/d/12906>.

[74] The text of this declaration is available at <https://www.regjeringen.no/en/aktuelt/nordic_declaration_solidarity/id637871/>.

[75] Buzan (note 40).

[76] Bailes, A. J. K., 'Societal security and small states', eds Archer, Bailes and Wivel (note 38).

Development Programme (UNDP) in 1994, covers economic, environmental, food and health dimensions as well as political, community and personal security conditions.[77] Applying these two perspectives to Arctic challenges adds further shades to the security picture, especially when all ten million of the wider Arctic's inhabitants, and not just the half a million or so classified as indigenous, are included.[78] Larger urban populations have their own social stresses and vulnerabilities, for instance to infrastructure breakdown.

Viewed in this framework, three additional issues arise. First, food security is a complex concern for people in non-agricultural or agriculturally marginal zones. Warming may boost local crops but cannot obviate high dependence on imports that could be disrupted by physical or economic factors. Meanwhile, the same climate processes are undermining traditional hunting and fishing, and methods of conserving food in cold and dry conditions. The West Nordic Council, an institution linking Iceland, the Faroe Islands and Greenland, made this its main policy research focus for 2014.[79]

Second, health (physical and mental) is already a problem in many Arctic locations for reasons including pollution and substance abuse as well as the difficulty of delivering welfare services. Warming will bring new pests and diseases, carried also by new waves of migration (see below) and tourism. The AC has studied the Arctic's special problems through its series of human development reports and has sponsored a health ministers' meeting on the subject, and the BEAC has sponsored cross-border health cooperation in North Western Europe, but responsibility rests at national level.[80]

---

[77] UN Development Programme, 'Human Development Report 1994: New Dimensions of Human Security', <http://hdr.undp.org/en/reports/global/hdr1994/>.

[78] Numbers/definitions are taken from the Arctic Human Development Report (ADHR), Nov. 2004, <http://svs.is/images/pdf_files/ahdr/English_version/AHDR_first_12pages.pdf>. An updated ADHR appeared in 2015, see <http://www.uarctic.org/news/2015/2/new-report-arctic-human-development-report-volume-ii-published/>.

[79] For further details see <http://www.vestnordisk.is/english/>. The United Nations explains food security at <http://www.un-foodsecurity.org/>.

[80] See Arctic Human Development Report (note 78). The Health Ministers meeting was held at Nuuk, Greenland, in Feb. 2011 during the Danish AC chairmanship. For further details of this cooperation see <http://www.beac.st/in-English/Barents-Euro-Arctic-Council/Working-Groups/Joint-Working-Groups/Health-and-Social-Issues>.

Third, the impact of population movements, especially when they make communities more multicultural, is a classic concern of societal security. Original populations may feel threatened by immigrants but the latter are also vulnerable, not least to unfamiliar climate conditions. In recent decades, the emigration of young people has been the main Arctic concern in areas reaching as far south as the Faroe Islands. Climate warming may require some internal resettlement, but also the temporary or longer-term implantation of foreign workers for projects where local labour is insufficient or underqualified. Antagonisms and security problems are not inevitable, if local society also perceives benefits, but possible dilemmas are illustrated by the case of Greenland. Reported plans to import thousands of Chinese labourers have caused dismay in local settlements, but the proposed solution of corralling them in camps is not ideal for their own human security.[81]

In solving problems of these kinds, private as well as public sector actions will be decisive, and consultation and the empowerment of people themselves is crucial. The need to consult, respect and protect indigenous peoples is recognized in all national and institutional Arctic strategies and is met to an extent through these communities' representation as Permanent Participants of the AC.[82] Less attention has been devoted to the role of non-indigenous Arctic residents—who have no special representative institutions—in security governance. Most national Arctic strategies so far have been promulgated on executive authority with no evidence of popular consultation, although Iceland and the Faroe Islands had theirs adopted by parliament. Among Arctic institutions, the BAEC has an extensive substructure allowing local authorities and social actors to be involved; the AC equivalents are limited to expert groups and a (large and diffuse) parliamentary body. More grassroots consultation is desirable not just on ecological, economic and societal risks but also on emergency management, where the

---

[81] 'Climate change brings new risks to Greenland, says PM Aleqa Hammond', *The Guardian*, 23 Jan. 2014.
[82] For further information about the Permanent Participants see <http://www.arcticcouncil.org/index.php/en/about-us/permanent-participants>. See also chapter 3 in this volume.

Nordic countries give much responsibility to local communities, and Iceland relies on a large volunteer rescue force (ICE-SAR).[83]

## VI. Multidimensional security and Arctic governance

The previous sections have discussed the wide range of security challenges facing the Arctic region and its inhabitants. The four categories presented here are overlapping and interdependent. Comprehensive, region-wide security management through a single governance system would seem called for, but is ruled out by the Arctic's well-entrenched national jurisdictions and the political characters of the nations concerned. None of the latter, except Denmark (mainland), Finland and Sweden, is ready even for EU-style partial surrender of sovereignty to a regional executive.

Does this matter? Armed conflict risks are discussed at length above, precisely to show that the supposed drivers behind them are questionable and the inhibitions substantial. General features of concerned states' interrelationships, notably nuclear deterrence, bolster the Arctic peace. Environmental, economic, societal and human risks flowing from climate change and its consequences are, meanwhile, immediate and rapidly growing. They may at any time generate major civil emergencies that local capacities, even pooled, are ill-prepared to master. Institutional frameworks already exist locally in North Western Europe, and region-wide in the AC, to tackle these challenges cooperatively. The AC's lack of military competence has not prevented it from inspiring cooperation among armed forces for civil ends, and its lack of direct legal competence has been overcome through case-by-case inter-governmental agreements.

Relying exclusively on the AC to fill remaining gaps in Arctic security governance would, however, be inappropriate as well as unrealistic. Its ambit is both too large and too small. The Arctic's subregions differ in political geography, climate dynamics, risk patterns and institutional capacity. The European fondness for overlapping, multilevel organizations is not shared elsewhere and may, for instance, be ill-suited to the Arctic's Pacific approaches where

---

[83] The Icelandic Association for Search and Rescue (ICE-SAR), <http://www.icesar.com/>.

traditional US–Russian and Russian–Chinese diplomacy could achieve much. Conversely, globalization makes the Arctic economy as well as environment highly dependent on outside powers. Environmental, financial, and sectorial regulation and enforcement are generally best provided through global mechanisms including the UN and its agencies. Regulation and self-regulation for private business even more clearly call for maximum inclusiveness. Applying universal norms of individual, community (including indigenous) and political rights may prove a better long-term solution for societal and human challenges than either leaving things to national discretion, or singularizing the Arctic as such.

Viewed this way, the toolbox for handling the Arctic's multi-dimensional security challenges is already well stocked. Applying such diverse institutional and procedural methods would not be considered unusual in other regions and, in Europe, is even commended under the notion of 'mutually reinforcing institutions'.[84] The Arctic case may seem more confusing than some because so many of the institutional actors involved are based outside the region or (like NATO) over the horizon. It could equally well be viewed as a model of 'messy' governance that reflects, and perhaps fits, the 21st century globalized environment. The messiness accommodates the need to find non-traditional solutions for non-traditional factors—from the power of multinationals to citizens' activism—and to let traditional and non-traditional players shape these solutions together in fields not yet circumscribed by international jurisprudence. This leaves space to develop and adapt in the face of further, as yet imponderable, Arctic changes. Perhaps not least important in political terms, it also allows the Arctic states to play different games, under different facets of their identity, on different issues when it suits them.

The Arctic region is far from alone in such messiness, and the introduction to this chapter warned against the temptation to singularize and isolate it. Doing so risks, on the one hand, overestimating the danger of conflict originating there, and on the other hand, missing the importance *and* availability of wider frameworks for addressing its problems. The Arctic is more like

---

[84] Defined in the OSCE Platform for Cooperative Security, 1999, <http://www.osce.org/mc/17562>.

than unlike other regions in the rapidity of change, the complexity of its security agenda, the messiness of relevant governance arrangements and also the continued centrality of state-to-state relations. Only Europe has gone some way towards overcoming the latter, and overly Eurocentric readings of the Arctic may miss the diversity both of challenges and management options across the whole region.

The jurisdiction held over Arctic territory, and most of the seas, by relatively strong states removes many security worries—urgent elsewhere—linked with state weakness, insurgency and terrorism. It creates its own questions, however, over how narrowly or widely Arctic 'ownership' should be drawn, in the strategic as well as regulatory and economic senses. Whether to give outsiders such as China, Japan and India observer status at the AC has lately been a contentious issue, resolved when they were admitted in 2013. This may prove a shrewd decision. Lacking voting rights, these powers so far lack openings to disturb Arctic stability unless through economic competition. The AC's eight member states may find common interest in curtailing undue interference, while educating the newcomers in emerging cooperative norms and making sure they pull their weight in climate mitigation. Rising powers' buy-in to positive Arctic initiatives should help especially when support from global bodies becomes vital.

These last judgements, admittedly, rely largely on the way in which low-key regional competition between the West and Russia, including in the Barents and Baltic regions, has managed to decouple itself from higher-level political confrontations in the past.[85] The crisis of 2014 in Ukraine came at a time when Arctic affairs were attracting unprecedented publicity, including speculation about clashes of interest and conflict risks. It was perhaps inevitable that the ensuing pattern of Western sanctions and Russian counteraction and recrimination would extend to the Arctic, affecting at least the military and some economic fields of cooperation. How lasting this impact will be, and whether the result is a slowdown of cooperative security building or a transition to actual

---

[85] Oldberg, I., 'The role of Russia in regional councils', Centre for Arctic Policy Studies Occasional Paper, Institute of International Affairs, University of Iceland, Reykjavik, 2014, <http://ams.hi.is/wp-content/uploads/2014/08/The-role-of-Russia_Online.pdf>.

risks of violence, remains to be seen. The establishment of the Arctic Coast Guard Forum by all eight AC states in October 2015 suggests that certain aspects of security cooperation with Russia will continue even though other channels have been suspended.[86] At any rate, these events strengthen the hypothesis that in the 21st century, conflict is more likely to enter the Arctic from the outside world than vice versa.

[86] 'Establishment of the Arctic Coast Guard Forum', Coast Guard Compass, 30 Oct. 2015, <http://coastguard.dodlive.mil/2015/10/establishment-of-the-arctic-coast-guard-forum/>.

# 3. Understanding national approaches to security in the Arctic

KRISTOFER BERGH AND EKATERINA KLIMENKO

## I. Introduction

Arctic security has become an important issue for several of the Arctic states. Compared to the relative indifference during the first decade and a half following the end of the cold war, references to the Arctic now appear in the defence and security doctrines of most states in the region. The approaches to defining security, security concerns and threats in the region vary from country to country: some emphasize the importance of environmental security, while others work hard to promote economic development or indigenous rights in the region. There are several factors that influence the development of national Arctic policies, including domestic politics and external factors such as climate change and fluctuations in the international arena. Different interest groups and actors, such as policy makers, military planners and environmental experts, may also perceive Arctic security differently, even within a country.

As argued by Alyson Bailes, there are several ways of perceiving 'Arctic security' and different state conceptions lead to diverse state strategies for dealing with the issues.[1] The strategies that the Arctic states employ to increase their perceived security in the region are largely implemented at the national level, often involving the armed forces, coastguard and other government functions. Wider international commitments and cooperation may also be reflected in national Arctic security concerns and interests.

This chapter explores how each of the eight Arctic states perceives its security interests in the region and how it works to enhance them.[2] The states are divided into three regions based on

---

[1] See chapter 2 in this volume.
[2] Referring to the 8 states with territory or maritime territory above the Arctic Circle and which are members of the Arctic Council: Canada, Denmark, Finland, Iceland, Norway, Russia, Sweden and the United States.

geography: North America (section II), Russia (section III), which for reasons explained below stands alone, and the Nordic countries (section IV). Section V then discusses to what extent the differences in approach influence prospects for a circumpolar security architecture or governance in the Arctic, and section VI provides some final conclusions.

## II. North America

The two North American and Arctic countries, Canada and the United States, share the world's longest unfortified border and a deeply integrated defence structure—not least through the North American Aerospace Defense Command (NORAD), a bilateral military command whose authority extends into the Arctic.[3] NORAD was established in 1958 and aims to provide aerospace warning, air sovereignty and defence for the two allies. Since 2006 it has also been responsible for maritime warning. In the USA, NORAD is administered by the US Northern Command (USNORTHCOM) and, together with the US European Command (USEUCOM), it forms the unified combatant command with Arctic responsibilities. USNORTHCOM is responsible for strategic planning for the Arctic.

While defence cooperation comes naturally to North America, political cooperation on the Arctic does not seem to be as straightforward. Canada and the USA have opposing views on a number of important issues, including disagreements over maritime territory in the region, the status of the Northwest Passage and an unresolved territorial disagreement in the Beaufort Sea.[4] These are outstanding issues, which could be dealt with relatively easily, yet they continue to be a distraction in US–Canadian relations in the Arctic. The territorial deal on the Barents Sea reached by Russia and Norway demonstrates that these issues are solvable and that solutions are greatly beneficial to international Arctic cooperation. However, although Canada and the USA are chairing the Arctic

---

[3] For further information see NORAD's website: <http://www.norad.mil/>.
[4] Canada claims that the Northwest Passage goes through internal waters whereas the USA claims that it is an international strait. For more detailed discussion see Bergh (note 7).

Council (AC) back to back—Canada (2013–15) and the USA (2015–17)—they have made little effort to coordinate their policies.

## Canada

Canadian Arctic policy is formulated domestically in its Northern Strategy, which covers Canada's three northern territories, and internationally in the Statement on Canada's Arctic Foreign Policy.[5] Both the domestic and international policy are based on four pillars: (*a*) exercising Arctic sovereignty; (*b*) promoting social and economic development; (*c*) protecting the environmental heritage; and (*d*) improving and devolving Northern governance. Canada's emphasis on sovereignty sets its policy apart from those of other Arctic countries, which tend to focus more on international cooperation.

In 2007 Canada took a very assertive stance on the Arctic and announced several large military investments in order to strengthen its sovereignty in the region.[6] Traditional 'hard' security issues seemed to be the main driver of that policy, which was a reaction to the region becoming more accessible to human activity because of melting ice.[7] Today, many of the previously announced investments are being scaled down and instead the focus is almost exclusively on the economic development of the region. Canada's three northern territories play an important role in Canadian national identity and they are more politicized than elsewhere in the country. Within the context of this domestic political debate, the Arctic is sometimes used as a tool for gaining political support.

---

[5] Canadian Government, *Canada's Northern Strategy: Our North, Our Heritage, Our Future* (Minister of Public Works and Government Services Canada: Ottowa, 2009); and Canadian Ministry of Foreign Affairs, Trade and Development, 'Statement on Canada's Arctic foreign policy', <http://www.international.gc.ca/arctic-arctique/arctic_policy-canada-politique_arctique.aspx?lang=eng>.

[6] Harper, S., Prime Minister of Canada, 'Strengthening Canada's Arctic sovereignty—constructing Arctic offshore patrol ships', 9 July 2007, <http://pm.gc.ca/eng/news/2007/07/09/strengthening-canadas-arctic-sovereignty-constructing-arctic-offshore-patrol-ships#sthash.a1K2YPny.dpuf>.

[7] Bergh, K., 'The Arctic policies of Canada and the United States: domestic motives and international context', SIPRI Insights on Peace and Security no. 2012/1, July 2012, <http://books.sipri.org/product_info?c_product_id=446>.

Canada has used its chairmanship of the AC to prioritize the establishment of a circumpolar business council.[8] At the same time, environmental issues have been steadily pushed to the background, sparking fears that the Canadian chairmanship would scale back on the AC's traditional issues in favour of an aggressive development agenda. The Canadian Prime Minister, Stephen Harper, has not made himself known as a champion of the environment during his administration—in the Arctic or elsewhere. Breaking with tradition, the news that the Canadian Chair of the AC would be the Minister of Health rather than the Minister of Foreign Affairs, and that the Chair of the Senior Arctic Officials (SAO) would be the President of the Canadian Northern Economic Development Agency instead of a senior diplomat, increased fears that Canada was trying to shift the focus of the Council. However, in June 2014, halfway through the Canadian chairmanship, the Chair of the SAO was replaced with a diplomat from the Department of Foreign Affairs, Trade and Development—a testament, perhaps, to the importance of diplomacy in the AC. In the wake of the Russian annexation of Crimea, that diplomacy was put to the test and it seems that Canada has navigated the crisis quite successfully, keeping the AC open for business in a challenging diplomatic environment.

However, the creation of an Arctic Economic Council (AEC) during the Canadian chairmanship sparked controversy as the AEC was accused of functioning as a lobby organization for 'big business' located outside the Arctic. As the AEC is set up to function independently of the AC, there is also concern that it might undermine the AC.[9]

Canada ratified the United Nations Convention on the Law of the Sea (UNCLOS) in December 2003 and therefore had to make its continental shelf claim no later than at the end of 2013. The Canadian Government has prioritized the mapping of the continental shelf north of Canada over the past ten years and the

---

[8] Canadian Ministry of Foreign Affairs, Trade and Development, 'Canada's Arctic Council chairmanship', <http://www.international.gc.ca/arctic-arctique/chairmanship-presidence.aspx?lang=eng>.

[9] Axworthy L. and Simon, M., 'Is Canada undermining the Arctic Council?', *Globe and Mail*, 4 Mar. 2015, <http://www.theglobeandmail.com/globe-debate/is-canada-undermining-the-arctic-council/article23273276/>.

scientific work has been carried out in cooperation with a number of other countries.[10] Surprisingly, the partial submission made in December 2013 was reduced to the Canadian claim in the Atlantic Ocean, but postponed the submission with regard to the Arctic Ocean, including areas around the North Pole, to a later date.[11] As a result, the partial claim of 2013 satisfied the 10-year deadline, but the timetable for Canada's final claim is still unclear.

Canada's aim remains to assert sovereignty in the Arctic. Nationally, it does this by way of the Canadian Armed Forces, including the Canadian Rangers, which have receive increased funding and have been given a stronger mandate to focus on Canadian sovereignty, rather than international operations. Internationally, Canada supports multilateral defence cooperation and has initiated the first-ever meeting between all the Arctic Chiefs of Defence (CHOD). However, Canada is against a stronger North Atlantic Treaty Organization (NATO) presence in the region, despite being a member, and has blocked discussions on the Arctic within the organization. Similarly, it has been reluctant to expand the number of permanent participants in the AC, clearly wanting to keep the circle of Arctic partners as small as possible.[12]

## The United States

Historically, the USA has not been that interested in the Arctic, not even from a defence perspective. However, in 2009 there was a presidential directive that put some emphasis on the region with a 'hard' security approach, but it lead to little in terms of increased capacity.[13] Today, the region is higher on the political agenda and the USA takes its commitments towards the AC seriously, not least because of the perceived increased geopolitical importance of the

---

[10] Canadian Ministry of Foreign Affairs, Trade and Development, 'Canada's extended continental shelf', <http://www.international.gc.ca/arctic-arctique/continental/index.aspx?lang=eng>.
[11] Government of Canada, *Partial Submission of Canada to the Commission on the Limits of the Continental Shelf regarding its continental shelf in the Atlantic Ocean*, 2013, <http://www.un.org/depts/los/clcs_new/submissions_files/can70_13/es_can_en.pdf>.
[12] Bergh (note 7).
[13] White House, 'Arctic region policy', National Security Presidential Directive no. 66 and Homeland Security Presidential Directive no. 26, 9 Jan. 2009, <http://georgewbush-whitehouse.archives.gov/news/releases/2009/01/20090112-3.html>.

Arctic and the personal interest in the region shown by policymakers like Hillary Clinton and John Kerry. In 2013 President Barack Obama announced a new 'National Strategy for the Arctic Region'. The strategy contains three 'lines of effort' to: (*a*) advance US security interests; (*b*) pursue responsible Arctic region stewardship; and (*c*) strengthen international cooperation.[14] While the strategy reaffirmed US commitment to the region, it did not include a budget or implementation plan. In 2015 President Obama issued an executive order aiming to coordinate the country's Arctic efforts in a more effective way.[15]

Since Alaska is the only US state with Arctic territory, much of the domestic Arctic debate becomes an issue of state versus federal authority—and domestic Arctic politics are largely synonymous with Alaskan state politics. The two senators from Alaska are outspoken supporters of a stronger US Arctic policy and increased federal spending, not least because it would benefit the state. From the Alaskan perspective, there is a perceived conflict between the focus on climate change on the one hand and economic development on the other, and the US chairmanship will have to carefully navigate the domestic aspects of this.

US defence interests in the Arctic are formulated in the 2013 'Arctic Strategy' published by the Department of Defense. The desired end state for the Arctic is described as 'a secure and stable region where US national interests are safeguarded, the US homeland is protected, and nations work cooperatively to address challenges'.[16] The strategy puts much emphasis on international cooperation in the Arctic and, accordingly, the US military takes part in international exercises and meetings on the region. The US attitude towards NATO in the region is unclear (the organization is not even mentioned in the strategy), but it seems that, in the current US budget environment, the tendency is towards burden sharing with allies.

---

[14] White House, *National Strategy for the Arctic Region* (White House: Washington, DC, 10 May 2003).

[15] White House, *Executive Order—Enhancing Coordination of National Efforts in the Arctic* (White House: Washington DC, 21 Jan. 2015).

[16] US Department of Defense, 'Arctic Strategy', 1 Nov. 2013, <http://thehill.com/sites/default/files/arctic_0.pdf>.

The US Navy, and perhaps primarily the Coast Guard, will be operating more in the region as it becomes more accessible and its main tasks are expected to be of a constabulary, rather than a military, nature.[17] In February 2014 the US Navy updated the 2009 Navy Arctic Roadmap and the new document sets out its operations from short-, medium- and long-term perspectives. According to the updated roadmap, the Arctic is 'is expected to remain a low threat security environment where nations resolve differences peacefully', but the navy is expected to play an increased role in the region and it should plan and budget accordingly.[18]

Overall, however, economic opportunity, rather than military concerns, has been the prominent driver of US policy in the Arctic.[19] Arguably, the most prestigious project in the US Arctic was the energy and petrochemical group Shell's offshore operation in the Chukchi Sea. Shell tried to get its project operational over several summers but, due to delays caused by safety concerns and protests from environmentalists, at the beginning of 2014 it announced that no activity would take place that year either. In September 2015, however, Shell announced that its drilling operations off Alaska would stop for the 'foreseeable future' as the drilling found little oil and gas.[20]

A 2014 report by the US General Accounting Office also downplays the impact of Arctic shipping on the US economy. It highlights the many problems with shipping in the US part of the Arctic, including 'geography, extreme weather and hard-to-predict ice floes'. Further, the report states that interest from the cruise ship industry is likely to be limited as the US Arctic offers little in terms of varying scenery and interesting ports.[21]

The Obama administration has continued to push for the ratification of UNCLOS but has repeatedly failed to bring the convention before the US Senate. This has excluded the USA from the major campaign of other Arctic states, which has been to enlarge

---

[17] Bergh (note 7).
[18] US Navy, 'The United States Navy Arctic Roadmap for 2014 to 2030', Feb. 2014, <http://www.navy.mil/docs/USN_arctic_roadmap.pdf>.
[19] Bergh (note 7).
[20] Macalister, T., 'Shell abandons Alaska Arctic drilling' *The Guardian*, 28 Sep. 2015.
[21] US Government Accountability Office (GAO), *Maritime Infrastructure: Key Issues Related to Commercial Activity in the US Arctic over the Next Decade*, GAO-14-299 (GAO: Washington, DC, Mar. 2014).

their Arctic territories by submitting claims on the continental shelf through UNCLOS. Though most maritime observers in the USA agree that it would be in the country's strategic interest to fully participate, domestic politics continue to keep it on the outside.[22]

The USA's AC chairmanship started in 2015. In July 2014 Admiral Robert J. Papp Jr. of the US Coast Guard became the US Special Representative for the Arctic, suggesting that maritime issues will be important to the AC's agenda in the coming two years. Indeed, Arctic Ocean safety, security and stewardship is one of the priorities of the US chairmanship, and as an example, there is an ambition to hold a full-scale Arctic search-and-rescue (SAR) exercise in 2016. Another priority will be improving economic and living conditions in the region, raising such issues as fresh water security, suicide prevention and improving telecommunications infrastructure. The final thematic priority for the US chairmanship is addressing the impacts of climate change. Here, the USA is following on from the Swedish chairmanship, which focused on black carbon and adaptation and resilience. There are also initiatives to enhance Arctic climate science.[23] The US agenda ahead is ambitious and a key factor to success will be to get Alaskans on board.

## III. Russia

### The strategic importance of the Arctic

The Arctic has traditionally been a zone of special interest for Russia, especially in terms of security. According to President Vladimir Putin, the 'Arctic is a concentration of practically all aspects of national security—military, political, economic, technological, environmental and that of resources'.[24] Indeed, Russia's Arctic zone is often seen as the country's main 'treasure chest',

---

[22] For more discussion of the USA's inability to ratify UNCLOS see Bergh (note 7).
[23] US Department of State, *Virtual Stakeholder Outreach Forum*, 2 Dec 2014, <https://www.arctic.gov/publications/presentations/Arctic_Council/US_Chairmanship_for_stakeholders.pdf>.
[24] Putin, V., Speech at the Meeting of the Security Council on state policy in the Arctic, Moscow, 22 Apr. 2014, <http://eng.kremlin.ru/news/7065>.

containing up to 95 per cent of Russia's gas and 60 per cent of Russia's oil reserves.[25] Russia's comprehensive 2008 Arctic strategy, 'Foundations of the Russian Policy in the Arctic region until 2020 and beyond', states that the Arctic zone is 'the strategic resource base of the Russian Federation', ensuring the socioeconomic development of the country.[26] The strategy also underlines the development of these resources as a primary goal of Russian policy in the region.

The strategic importance of the Arctic region is also closely associated with the development of the Northern Sea Route (NSR). The 'Russian Transport Strategy up to 2030' underlines the significance of the NSR with regard to its role in the supply of the northern regions of the country, the so-called *severny zavoz* (north delivery).[27] Russia also plans to transform the NSR into 'an international transport artery capable of competing with traditional sea routes in costs, safety and quality'.[28] The most ambitious prognosis states that by 2030 the turnover of the NSR will be 64 million tonnes.[29]

Further, the Arctic is seen as a very important element of Russia's military capability. According to the 2015 Maritime Doctrine of the Russian Federation, the military significance of the Arctic is 'determined by the particular importance of providing free access of the Russian fleet to the Atlantic' and 'the crucial role of the Northern Fleet for the defence of the state in marine and ocean areas'.[30] The Northern Fleet remains Russia's biggest fleet and it is also an important part of the country's nuclear deterrence forces.

---

[25] Pushhaev, J., 'Poljus pretknovenija Jeksperty obsuzhdajut vozmozhnost' vozniknovenija "gorjachih tochek" v Arktike' [Pole block: Experts discuss the possibility of 'hot spots' in the Arctic], *Rossiyskaya Gazeta*, 17 May 2011, <http://www.rg.ru/2011/05/17/polus.html>.

[26] Foundations of the Russian Policy in the Arctic region until 2020 and beyond, <http://www.scrf.gov.ru/documents/98.html>.

[27] Transport Strategy of the Russian Federation for the period until 2030, <http://rosavtodor.ru/storage/b/2014/06/24/trans_strat.pdf>.

[28] Председатель Правительства Российской Федерации В.В.Путин принял участие во втором Международном арктическом форуме «Арктика – территория диалога» [Prime Minister Vladimir Putin takes part in the second International Arctic Forum, 'The Arctic—Territory of Dialogue'], 22 Sep. 2011, <http://archive.government.ru/special/docs/16536/>.

[29] 'Syevyerniy morskoy put' oovyelichit groozooborot v 50 raz' [Northern Sea Route increases turnover by 50 times], *Izvestia*, 4 Aug. 2011, <http://izvestia.ru/news/496692>.

[30] *Maritime Doctrine of the Russian Federation*, 2015, <http://static.kremlin.ru/media/events/files/ru/uAFi5nvux2twaqjftS5yrIZUVTJan77L.pdf> (in Russian).

In addition to the goals clearly stated in these major policy documents, there is another element that explains Russia's strategic interest in the Arctic: Russia's 'great power' image. Three key components of Russia's Arctic policy—energy, the NSR, and the northern military and naval capabilities—are of crucial importance to Russia's image of itself as a 'great power' internationally.

The Arctic is an important element of the Russian leadership's domestic policy as well. The 'great power agenda' has been used as proof, for the domestic audience, of the effectiveness and success of President Putin's regime.[31] Thus, the successful development of the Arctic zone, defending Russia's rights to the Arctic shelf and continuing scientific research in the Arctic are all important elements of Putin's claims of success in the revival of Russia as whole and the Russian North in particular. They demonstrate that the country is back in the international arena and can defend its national interests. For instance, in 2007 there was a famous Russian expedition to the North Pole, during which the Russian flag was planted on the seabed of the Arctic Ocean. The act symbolized Russia's return to the Arctic and was mostly directed at the domestic audience.[32]

### Threat assessment

Although the process of forming Russian Arctic policy is very centralized, and it might seem that Russia only has one view on Arctic affairs, in reality the situation is more complicated and a number of state and non-state actors are involved in its determination and implementation. Putin's administration and the Russian Security Council decide core policy, underlining the strategic importance of the Arctic region, although other state agencies are involved at different levels and in their respective fields. For example, the Ministry of Natural Resources and Ecology and the Ministry of Energy are important players within the development of Arctic resources, while the Ministry of Transport is supervising the development of the NSR. The state companies Gazprom and

---

[31] Laruelle, M., *Russia's Arctic Strategies and the Future of the Far North* (M.E. Sharpe: Armonk, NY, 2014), p. 9.
[32] Baev, P., *Russia's Race for the Arctic and the New Geopolitics of the North Pole*, Jamestown Occasional Paper (Jamestown Foundation: Washington, DC, Oct. 2007).

Rosneft also play a very significant role as they are directly involved in the development of Arctic resources, as do some private companies (e.g. Novatek) that are working in the Arctic.

Different institutions view security threats in the Arctic differently and each has its own approach to dealing with them. The Russian Security Council and other defence-oriented institutions see the protection of sovereignty and improving military and other aspects of security as the highest priorities.[33] They also view threats in the region from a very 'hard' security-oriented perspective. This 'security first' approach was dominant in the first years of Russia's return to the Arctic and remains strong today.

Another approach concentrates on the need to develop the region's economic potential, primarily the development of natural resources and shipping routes, and emphasizes international cooperation as one of the main ways of realizing this. Economic actors, such as the Ministry of Natural Resources and Ecology, Gazprom and Rosneft, mostly support this 'pro-cooperation' approach, which has become more influential since 2010.[34]

According to the Secretary of the Russian Security Council and one of the main supporters of the security first approach, Nikolai Patrushev, the main objective of Russian activities in the Arctic is to avoid having Russia's sovereign rights in the region undermined by other states.[35] In fact, the protection of sovereignty is seen as one of the top priorities of Russia's Arctic policy and the violation of Russia's sovereign rights is seen as one of the major security concerns.

According to Patrushev, among the major challenges that Russia faces are the aspirations of other Arctic players—states, companies and individuals—that could result in disputes and conflicts: 'There is the possibility of the exacerbation of the international relations in the Arctic around energy and bio-resources, fresh water as well as access to the transport routes both maritime and air.'[36] He also

---

[33] Laruelle (note 31), p. 9.
[34] Laruelle (note 31), p. 7.
[35] Interview with Secretary of the Russian Security Council Nikolai Patrushev: 'Arctic war. Why Russian ice to protect their borders', *Argumenty i Facty*, 10 Apr. 2013, <http://www.scrf.gov.ru/news/22/778.html>.
[36] Объем перевозок по Севморпути с августа 2011 года вырос на 20% [The volume of traffic on the Northern Sea Route increased by 20% in August 2011], 12 Apr. 2012

noted the increasing military activities of Arctic states, primarily the USA, Canada and Denmark.[37]

However, with regard to threats coming from other states, perhaps the main aggravator in the Arctic region for Russia is NATO. In 2010 President Dmitry Medvedev stated that Russia is watching NATO's increased activity in the Arctic 'intently and with some concern' and that it 'could do without NATO in the Arctic because it is part of our common heritage, which, strictly speaking, does not have anything to do with military objectives. We are fully capable of managing there with the use of economic regulation and international agreements we sign'.[38] The Foreign Minister, Sergei Lavrov, also stated that Russia did not see what benefit NATO could bring to the region: 'I do not think that NATO will do the right thing by taking it upon itself to determine, who will and how to decide issues in the Arctic.'[39]

Finally, it is also worth mentioning that since the Arctic region is, to a large extent, part of Russia's broader security agenda, a significant part of the Arctic strategy aims to both maintain the strategic balance with the USA and protect Russia's international image. The negative reaction to NATO and US actions in the Arctic is part and parcel of the larger relationship between Russia and NATO, and not about the Arctic region in particular.

### Russia's military capabilities in the Arctic

Restoring military capabilities in the Arctic has become a significant element of ensuring sovereignty. Two strategy documents— the 'Foundations of the Arctic Policy until 2020 and beyond' and 'Development of the Arctic Zone of the Russian Federation and National Security Efforts for the period up to 2020'—aim to keep

<http://pripolyarie.ru/news/2012/04/obem-perevozok-po-sevmorputi-s-augusta-2011-goda-vyros-na-20-patrushev.htm>.

[37] Interview with Patrushev (note 35).

[38] Medvedev, D., 'News conference following Russian–Norwegian talks', Transcript, 15 Sep. 2010, <http://eng.kremlin.ru/transcripts/946>.

[39] Russian Ministry of Foreign Affairs, 'Transcript of Russian Foreign Minister Sergey Lavrov's Remarks and Answers to Media Questions at Joint Press Conference Following Talks with Canadian Foreign Minister Lawrence Cannon, Moscow, September 16 2010', <http://www.mid.ru/bdomp/Brp_4.nsf/arh/1EC648EE265E5A50C32577A00058E338?OpenDocument>.

up the 'favourable operational regime' and maintain 'the required combat potential', by creating special military groupings under the armed forces and strengthening border control systems.[40] In 2011 Russia announced plans to establish two Arctic brigades, the first of which was established in January 2015 in the city of Alakurtti, 60 kilometres from the Finnish border.[41] By 2016 another brigade will be established on the Yamal peninsula.[42]

As part of the overall rearmament programme for the Russian military, an ambitious rearmament programme was also launched for the Northern Fleet. For example, Russia is replacing the Soviet-era nuclear-powered ballistic missile submarines (SSBNs) with new Borei-class ones; in total, Russia is planning to build 8 Borei-class submarines that are to be distributed between the Pacific and Northern fleets. The first of them, *Yurik Dolgoruky*, joined the Northern fleet in 2013. It is also planned to build 8 Yasen-class attack submarines. The first of them, *Severodvinsk*, entered into service with the Northern Fleet in 2013. Within the overall rearmament programme, Russia is planning to build 51 surface ships, including up to 15 frigates and 25 corvettes.[43] According to the Commander of the Northern Fleet, Admiral Vladimir Korolev, 2 landing ships, 1 destroyer, 5 frigates and 5 mine-hunters will be allocated to the Northern Fleet by the end of 2020.[44]

Recently Russia has restored a number of airfields beyond the Arctic Circle and the military base on the New Siberian Islands.[45]

---

[40] Development of the Arctic Zone of the Russian Federation and National Security Efforts for the period up to 2020, <http://minec.gov-murman.ru/files/Strategy_azrf.pdf>; and Foundations of the Russian policy in the Arctic region until 2020 and beyond, <http://www.scrf.gov.ru/documents/98.html>.

[41] Первая в России арктическая мотострелковая бригада приступила к боевой подготовке [Russia's first Arctic motorized infantry brigade started combat training], 1st Channel, 18 Jan. 2015, <http://www.1tv.ru/news/social/275906>.

[42] Pettersen, T., 'Shoygu to visit Arctic Brigade in Alakurtti', *Barents Observer*, 8 May 2015, <http://barentsobserver.com/en/security/2015/05/shoygu-visit-arctic-brigade-alakurtti-08-05>.

[43] Federov, Y., Государственная Программа Вооружений–2020: Власть И Промышленность [State Rearmaments Programme 2020: power and industry], Index bezopasnosty no. 4 (107), 2013 <http://www.pircenter.org/media/content/files/12/13880454280.pdf>.

[44] Северный флот получит более 40 кораблей до 2020 года [Northern Fleet will receive more than 40 ships until 2020], Kommersant, 8 Apr. 2014, <http://www.kommersant.ru/doc/2447988>.

[45] Federal Security Service Board meeting, 7 Apr. 2014 <http://kremlin.ru/news/20724>.

The head of the National Center of Defense Management, Mikheil Mizincev, stated that in total 13 airfields, 1 air-force test range, and 10 radar sites and direction centres would be opened in the Arctic in the coming years.[46] In December 2014, the Ministry of Defence established the Joint Military Command 'North' (OSK Sever) for the Arctic region on the basis of the Northern Fleet.[47] Interestingly, Russian officials do not see these actions and increases in military capabilities as a militarization of the region but rather as 'strengthening and exercising state sovereignty'.[48]

As well as the traditional security threats, Russian Arctic policy documents and Russian officials emphasize that climate change and increasing economic activity in the region have brought a number of new security challenges. Russia has the longest maritime border in the region, which is now largely unprotected due to receding ice and makes the region vulnerable to terrorism, organized crime and illegal migration. The active functioning of the NSR raises a number of concerns related to ensuring the safety of navigation. There is also the danger of terrorist attacks on oil platforms and other important infrastructure.[49]

In order to strengthen the security of its northern borders and the safety of transportation routes and infrastructure, Russia plans to build 10 Arctic SAR centres and establish 20 new Arctic border posts along the NSR.[50] The first 3 centres have opened in Naryan-Mar, Dudinka and Arkhangels. Additionally, 4 regional SAR teams and fire-rescue units of various departments are now operational. The region has 2 maritime rescue coordination centres, in Murmansk and Dickson; 3 marine rescue sub-centres, in Arkhangelsk,

---

[46] Russian army beefs up Arctic presence over Western threat, Russia Today, 29 Oct. 2014, <http://rt.com/news/200419-russia-military-bases-arctic/>.

[47] В Арктике начинает действовать объединенное стратегическое командование 'Север' [Joint Military Command becomes active in the Arctic], Itar-Tass, 1 Dec. 2014, <http://itar-tass.com/spb-news/1612884>.

[48] Vasiliev, A., 'Security and cooperation in the Arctic: new factors, challenges and prerequisites', Report from the international conference 'Security and Cooperation in the Arctic New Frontiers', Murmansk, 12 Apr. 2012, *Arctic Herald*, no. 2 (2012); and Pettersen, T., 'Arctic Ambassador: there is no militarization of the Arctic', *Barents Observer*, 3 Feb. 2014, <http://barentsobserver.com/en/arctic/2014/02/arctic-ambassador-there-no-militarization-arctic-03-02>.

[49] Pettersen (note 48).

[50] 'Rossiya zadyelayet arktichyeskiye «diri»' [Russia patches up Arctic holes], Izvestia, 16 Apr. 2012, <http://izvestia.ru/news/522020>.

Tiksi and Pevek; and bases for rescue property and equipment for the liquidation of oil spills, in Dixon, Tiksi, Pevek and Provideniya.[51]

Russia is also planning to reinforce air support for emergency SAR operations in the region.[52] In addition, Russia will create several infrastructure hubs in the Arctic, to be used as temporary stations for Russian warships and border guard vessels in remote areas of the Arctic seas.[53]

## Policy shifts

During the early years of its Arctic policy revival, Russia put major emphasis on security aspects and the 'security first' agenda largely dominated. In 2010, however, it took a significant shift towards cooperation. At the 2010 forum, 'Arctic: the Territory of Dialogue', Putin took a major step towards de-securitizing the Arctic agenda, promoting the idea of cooperation in the Arctic and improving Russia's image in the region. He stated that 'the existing issues in the Arctic, including those related to the continental shelf, can be resolved in a spirit of partnership through negotiations and on the basis of existing international law' and that 'preserving the Arctic as a zone of peace and cooperation is of the utmost importance'. He also claimed that speculation regarding conflict in the Arctic 'lacks real grounds'.[54]

A number of factors contributed to this shift. First, the existing legal framework (UNCLOS) has been beneficial in the process of determining the limits of the Arctic shelf. Second, the increasing attention of non-Arctic states towards the region has led Russia to team up with other Arctic states, developing 'the rules of the game'

---

[51] [The main directions of Ministry of Emergency Situations of Russia's activities in the Arctic], 9 June 2014, <http://www.mchs.gov.ru/dop/info/smi/news/item/989168/>.

[52] 'Dopolnityel'naya aviagrooppa poyavitsya v arktichyeskoy zonye RF na sloochay ChS' [Additional air grouping will be established in the Arctic Zone of the Russian Federation], RIA Novosti, 14 Mar. 2012, <http://ria.ru/arctic_news/20120314/594840472.html>.

[53] 'Russia to Set Up Naval Infrastructure in Arctic – Patrushev', Sputnik, 6 Aug. 2012, <http://sputniknews.com/military/20120806/175015455.html>.

[54] Речь В.В. Путина I Международном Арктическом форуме [Speech by Vladimir Putin at the First International Arctic Forum], Russian Geographic Society, 23 Sep. 2010, <http://www.rgo.ru/ru/page/rechi-vv-putina-i-sk-shoygu-na-i-mezhdunarodnom-arktic heskom-forume>.

in a closed circle. Third, the economic development of Russian Arctic resources has proven to be difficult without foreign investment and technology; by scaling down the conflict rhetoric, Russia has created fertile ground for technology exchange and investment in its Arctic projects.

Russia has once again confirmed its adherence to the procedures of UNCLOS and put significant efforts into the delimitation of its borders with Arctic neighbours as well as defining the limits of its Arctic shelf. It has carried out a number of expeditions to collect evidence for its application for the extension of the Russian continental shelf, within UNCLOS procedures. In 2013 Russia achieved its first success with a positive decision from the UN Commission regarding the Okhotsk Sea enclave.[55] In April 2014 Putin once again underlined the importance of delimitation, stating that 'a pressing issue that requires careful work is the legal formalisation, in line with international law, of the outer boundary of Russia's continental shelf in the Arctic Ocean'.[56]

Since 2010 Russian Arctic policy has placed the greatest emphasis on pursuing a cooperative approach. For instance, the Foreign Minister, Sergei Lavrov, has repeatedly underlined the importance of international cooperation in ensuring the success of Arctic development: 'Many of our national interests in the region can only be realized in close cooperation with the other Arctic states.'[57] While Russia's SAO, Anton Vasiliev, has denied that there would be conflict between the Arctic states even if the existing 'rules of the game' are not challenged.[58]

Within this doctrine, Russia has focused strongly on the activities of the AC and claims to be among those responsible for the

---

[55] 'Resource-rich Sea of Okhotsk all Russian', The Voice of Russia, 17 Mar. 2014, <http://voiceofrussia.com/news/2014_03_17/Resource-rich-Sea-of-Okhotsk-all-Russian-3729/>.

[56] Meeting of the Security Council on state policy in the Arctic, 22 Apr. 2014, <http://eng.kremlin.ru/transcripts/7065>.

[57] Статья Министра иностранных дел России С.В.Лаврова 'Нуукская Декларация: новый этап сотрудничества арктических государств', 'Арктика: экология и экономика', № 3, 2011 год [Article by Russian Foreign Minister Sergei Lavrov, 'Nuuk Declaration: A New Stage of Cooperation among Arctic States', Arktika: Ekologiya i Ekonomika Magazine, no. 3, 2011], <http://www.mid.ru/brp_4.nsf/0/5083A33779726F2BC32579280022719B>.

[58] Vasiliev (note 48).

adoption of the first legally binding documents signed within it.[59] Russian officials support the strengthening of the AC and its gradual transformation into a fully-fledged international organization.[60] They have also underlined the importance of military cooperation in the region—beyond the AC, since there is no military cooperation within that forum—particularly through Arctic CHOD Staff Meetings.[61]

On a bilateral level, in 2010, after 40 years of painful negotiation, Russia resolved its dispute with Norway over their maritime borders in the Barents Sea. In the same year, Russia and Norway also resumed joint naval exercises (Pomor) after a 16-year break.

The pro-cooperation approach, however, has not entirely ruled out the security first approach in the Arctic. Nor has it eliminated the fact that the Arctic is an important element of domestic politics—Russian officials are usually harsher in their statements when talking to the domestic audience, especially to military personnel. In international forums, for instance, Russian officials claim that there is no danger of militarization in the Arctic despite increased military capabilities in the region, but when appealing to the domestic audience they often criticize the Arctic states for their actions there. In February 2013 Putin talked to the expanded meeting of the Defence Ministry Board and raised concerns that 'there are attempts to undermine strategic balance and there is danger of militarization in the Arctic'.[62] Then, in October 2013, he pointed out that US submarines were still patrolling the Norwegian coast and that Russia should keep up its capabilities to respond to this kind of activity.[63] Thus, in order to demonstrate its power in the region to both domestic and international audiences, Russia continues to undertake various symbolic acts: for example, the expeditions of the nuclear-powered battle cruiser *Pyotr Veliky*

---

[59] Prime Minister Vladimir Putin takes part in the second International Arctic Forum (note 28).

[60] Lavrov (note 57).

[61] Vasiliev (note 48).

[62] 'Expanded meeting of the Defence Ministry Board', President of Russia News Release, 27 Feb. 2013, <http://en.kremlin.ru/events/president/news/17588>.

[63] Встреча с активом партии «Единая Россия» 3 октября 2013 года, Московская область [Meeting with activists of the party 'United Russia'], 3 Oct. 2013, <http://www.kremlin.ru/news/19356>.

along the NSR to the Novosibirskie Islands in the summers of 2012 and 2013.[64]

In the aftermath of the Ukraine crisis Russia has upped its security rhetoric and started to put more effort into strengthening military capacity in the Arctic. In December 2014 the Russian Defence Minister, Sergey Shoigu, underlined that a 'broad spectrum of potential threats to Russia's national security is now being formed in the Arctic'.[65]

The changing rhetoric has been reflected in key security documents (although in a much more cautious formulation). The task of 'protecting Russian interests in the Arctic' appeared for the first time in the Russian Military Doctrine (2014). The Doctrine, like its predecessor, cites NATO expansion closer to Russia's borders as the number one external military threat.[66]

New amendments to the Maritime Doctrine adopted in July 2015 have focused on two regions: the Atlantic and the Arctic. NATO's global activities are seen as the major security concern on the Atlantic side, while the significance of the Arctic is determined by the need to provide limitless access to the Atlantic and Pacific oceans, as well as the key importance of the military capabilities of the Northern Fleet for the defence of Russia. Additionally, the updated version of the Maritime Doctrine has underlined that its main goal of 'lowering the threats in the Arctic region' will be achieved by, among other things, strengthening the Northern Fleet.[67]

In summary, Russia's security concerns and objectives in the Arctic can be divided into three main areas: (*a*) the protection of sovereignty and economic interests in the region; (*b*) the changing environment and new security challenges emerging as a result of climate change; and (*c*) the Arctic as a part of Russia's larger defence capabilities, including maintaining the nuclear strategic balance with its main perceived competitor, the USA/NATO.

---

[64] 'Russian military resumes permanent Arctic presence', Russia Today, 14 Sep. 2013, <http://rt.com/news/russian-arctic-navy-restitution-863/>.
[65] 'Russia to build network of modern naval bases in Arctic—Putin', Sputnik International, 14 Apr. 2014, <http://sputniknews.com/military/20140422/189313169.html>.
[66] Military Doctrine of the Russian Federation 2014, <http://www.rg.ru/2014/12/30/doktrina-dok.html>.
[67] Maritime Doctrine of the Russian Federation 2015, President of Russia News Release, <http://kremlin.ru/events/president/news/50060>.

Although Russia's attempts to strengthen its sovereignty and defend its economic interests in the region are sometimes seen by other Arctic states as overly assertive, the results have actually been beneficial for Arctic security and cooperation. The country's support for UNCLOS and its mechanisms of delimitation have significantly de-escalated the conflict rhetoric in the region, as has its active engagement with the AC regarding SAR and oil spill agreements. According to a number of Western officials and diplomats, Russia has become a very constructive player in the Arctic and has put significant effort into engaging with other Arctic states.[68]

## IV. The Nordic countries

The Nordic countries, Denmark, Finland, Iceland, Norway and Sweden, have chosen different paths for their national defence policies. While some have joined the transatlantic defence community, others have remained neutral. Three of the five have joined the European Union (EU), but Norway and Iceland have chosen to remain outside it. Despite these differences, there is a strong sense of community in the Nordic region and deeper defence integration is always a topic of debate in these countries. A Norwegian diplomat, Thorvald Stoltenberg, wrote a report in 2009 recommending increased cooperation and coordination on security and foreign policy between the Nordic countries and, while the region may not have reached the levels that he suggested, there has certainly been a trend and ambition towards increased cooperation.[69]

In 2011 the five Nordic countries came together on the last of Stoltenberg's recommendations and accepted a Nordic declaration of solidarity, stating that it is natural for the Nordic countries to cooperate in a spirit of solidarity when it comes to challenges in the foreign and security policy arena. More specifically, the declaration mentions challenges caused by natural and man-made

---

[68] 'Iceland's saga: a conversation with Ólafur Ragnar Grímsson', *Foreign Affairs*, Jan./Feb. 2014.
[69] Stoltenberg, T., 'Nordic cooperation on foreign and security policy: proposals presented to the extraordinary meeting of Nordic foreign ministers in Oslo on 9 February 2009', Norwegian Ministry of Foreign Affairs, 9 Feb. 2009, <http://www.regjeringen.no/en/dep/ud/Whats-new/news/2009/nordic_report.html?id=545258>.

disasters, cyber attacks and terrorist attacks.[70] Significantly, the declaration that was accepted did not include the militarily binding guarantees that Stoltenberg suggested—it is a declaration that they will act, but not on how.

Their solidarity also has an institutional home: the Nordic Defence Cooperation (NORDEFCO), which is a joint effort by the Nordic countries to cooperate in order to cut defence costs. NORDEFCO is not an alliance, but rather a way to explore synergies, increase effectiveness and improve coordination. The chairmanship is rotated between the five member states on an annual basis. Denmark, which held the chairmanship in 2012, made the Arctic a priority for NORDEFCO.[71]

In the countries that do not have an immediate proximity to the Arctic Ocean, such as Finland and Sweden, the changes in the Arctic become part of a much broader defence debate—which often centres around growing Russian assertiveness and the impacts that will have on national defence strategies. Russia's activities in the Arctic are used in national debates as motivation for increased defence spending and capacity building. Politically, however, both Finland and Sweden are working hard to tone down any suggestions of armed conflict in the region and are heavily committed to multilateralism and environmental issues there, not least through the AC.[72]

Finland, Norway and Sweden have an agreement on cross-border training that allows for joint exercises across the states' territories. In the summer of 2015 a large air-force exercise was hosted by the three countries, which included 115 aircraft and 3600 personnel from nine, mostly NATO, countries. The exercise scenario revolved

---

[70] Norwegian Ministry of Foreign Affairs, 'Enige om nordisk solidaritetserklæring' [Agreement on Nordic declaration of solidarity], Press release, 5 Apr. 2011, <http://www.regjeringen.no/nb/dep/ud/pressesenter/pressemeldinger/2011/norden_enige.html?id=637871>.

[71] Danish Ministry of Defence, 'Nordic Defence Cooperation (NORDEFCO)', <http://www.fmn.dk/eng/allabout/Pages/NordicDefenceCoorporation.aspx>.

[72] For more information see Finnish Prime Minister's Office, *Finland's Strategy for the Arctic Region 2013* (Prime Minister's Office: Helsinki, 2013), <http://vnk.fi/documents/10616/334509/Arktinen+strategia+2013+en.pdf/6b6fb723-40ec-4c17-b286-5b5910fbecf4>; and Swedish Ministry for Foreign Affairs, 'Sveriges strategi för den arktiska regionen' [Sweden's strategy for the Arctic region], <http://www.regeringen.se/contentassets/2c099049a492447b81829eb3f2b8033c/sveriges-strategi-for-den-arktiska-regionen>.

around a border dispute and natural resources.[73] The Swedish armed forces clearly stated that the aim of the exercise was to increase capacity as well as to send 'security policy signals'.[74] The Russian response was an exercise of their own, including 250 planes and 12 000 soldiers.[75]

## Denmark

Denmark has a complex relationship with the Arctic as its mainland is not located in the region but Greenland is: Arctic issues get intertwined with those of Greenlandic independence and economic sustainability. There is also a social and cultural aspect to Denmark's engagement in the region, not least because of the indigenous population of Greenland. Greenland is huge, with a relatively small population, and its defence and foreign policy is conducted from the remote Danish mainland. The limited capacity of the Danish military makes it difficult to routinely patrol borders and assert sovereignty over Denmark's entire Arctic territory. However, Danish and Greenlandic authorities do not see any conventional military threats in the region and their focus is elsewhere, notably in the spheres of social and human security.[76]

Similarly to Canada, Denmark has submitted a series of claims to the Commission on the Limits of the Continental Shelf (CLCS), arguing that the Danish continental shelf extends beyond 200 nautical miles of Greenland and the Faroe Islands. In December 2014 it submitted further claims in the Arctic, including the North Pole, suggesting that overlapping claims (with at least Canada) are likely.

Denmark expects the situation with potentially overlapping claims in the Arctic to be resolved according to the recommendations

---

[73] 'Arctic Challenge Exercise 2015', Swedish Armed Forces <http://www.forsvar smakten.se/sv/var-verksamhet/ovningar/avslutade-ovningar/arctic-challenge-exercise-2015/scenario-ace-2015/>.
[74] 'Amerikanska B-52:or sänds in över Sverige' [American B-52s deployed over Sweden], *Svenska Dagbladet*, 20 May 2015.
[75] 'Russia begins massive air force exercise', BBC News, 26 May 2015.
[76] For a discussion on societal and human security approaches in the Arctic see chapter 2, section V, in this volume.

of the CLCS.⁷⁷ In fact, it was the Danish Foreign Minister and the Greenlandic Premier who took the initiative of the Ilulissat conference in 2008, which led to a declaration stating that the five Arctic states would settle their territorial claims in the region according to international law.⁷⁸ Denmark has also had other disagreements in the Arctic, including over Hans Island, a tiny island in the Kennedy Channel, which both Denmark and Canada claim ownership of. However, it is expected that this disagreement will be solved diplomatically and does not affect the two countries' overall relationship.

## Finland

Both Sweden and Finland have strong traditions in polar research and, as a result of the yearly freezing of the Baltic Sea, they possess the word's greatest ice-breaking capabilities outside Russia. Finland's Arctic strategy (2013) states that the country subscribes to a comprehensive security concept in the Arctic, which means 'securing the vital functions of society through close cooperation between the authorities, industry, NGOs and citizens'.⁷⁹ The strategy states that military conflict in the region is 'improbable' but recognizes the benefits of the armed forces' capacity to operate in the region, not least when it comes to matters such as SAR.⁸⁰

Finland is quite unique in the Arctic context because of its outspoken support for the EU to take on a larger role in the region. Despite being in the company of many EU sceptics, Finland maintains support for 'the formulation of the EU's policy towards the Arctic and the reinforcement of its role in the region'.⁸¹ The EU's role in the Arctic is more controversial than that of other AC observers. In Canada there is a strong perception that the EU is not

---

⁷⁷ For more information on Denmark's continental shelf ambitions see Rigsfællesskabets Kontinentalsokkelprojekt [The Danish Realm's Continental Shelf Project], <http://a76.dk/>.

⁷⁸ 'The Ilulissat Declaration', Arctic Ocean Conference, Ilulissat, Greenland, 27–29 May 2008, <http://www.oceanlaw.org/downloads/arctic/Ilulissat_Declaration.pdf>.

⁷⁹ Finnish Prime Minister's Office, *Finland's Strategy for the Arctic Region 2013* (Prime Minister's Office: Helsinki, 2013), http://vnk.fi/documents/10616/334509/Arktinen+strategia+2013+en.pdf/6b6fb723-40ec-4c17-b286-5b5910fbecf4>, p. 40.

⁸⁰ Finnish Prime Minister's Office (note 79).

⁸¹ Finnish Prime Minister's Office (note 79), p. 47.

sensitive enough to traditional ways of life in the Arctic, specifically because of its scepticism regarding sealing and seal products. The EU also has the capacity to coordinate its members' Arctic policies, which is viewed with some concern by the non-EU members in the region.

## Iceland

In Iceland the increased accessibility of the Arctic is viewed as an economic opportunity and a way to increase interaction with global powers such as China. Icelandic President Ólafur Ragnar Grimsson has been a strong proponent of bringing China closer to Arctic cooperation processes. In 2013, before China and other non-Arctic states were formally accepted as observers in the AC, President Grimsson took the initiative of creating the Arctic Circle, a more inclusive organization than the AC that would, among other things, promote business interests in the region. The first Arctic Circle meeting was held in Reykjavik in October 2013 and follow-up meetings took place in 2014 and 2015.[82] The Icelandic Government does not necessarily share Grimsson's enthusiasm towards China and there is nothing specific about increasing cooperation with the country in Iceland's official Arctic strategy from 2011.[83] However, even if Grimsson's ideas are not fully anchored in Iceland's domestic politics, they have received much attention abroad. Iceland has also signed a free trade agreement with China, with the hope of increasing direct trade between the two countries, possibly using the NSR.[84]

Yet more immediate concerns for Iceland include its strong dissatisfaction with meetings held exclusively among the five Arctic littoral states (Iceland considers itself to be one) and challenges to fishing that might occur as the migratory patterns of fish change along with the climate. Iceland also strives to increase cooperation

---

[82] For more information see the Arctic Circle website, <http://www.arcticcircle.org/>.

[83] 'A parliamentary resolution on Iceland's Arctic policy', Approved by Althingi at the 139th legislative session, 28 Mar. 2011, <http://www.mfa.is/media/nordurlandaskrifstofa/A-Parliamentary-Resolution-on-ICE-Arctic-Policy-approved-by-Althingi.pdf>.

[84] Icelandic Ministry for Foreign Affairs, 'Free Trade Agreement between Iceland and China', 15 Apr. 2013, <http://www.mfa.is/foreign-policy/trade/free-trade-agreement-between-iceland-and-china/>.

with Greenland and the Faroe Islands in order to strengthen the international and economic position, as well as the politico-security dimension, of the West Nordic countries.[85]

Iceland does not have a navy and the Icelandic Coast Guard is expecting an increased workload in relation to SAR and unlawful fishing. Iceland has previously been heavily dependent on the USA for air surveillance but, after the USA withdrew from the Keflavik airbase, opportunities for other NATO countries and increased Nordic cooperation in this regard have emerged. Responsibility for the surveillance of Iceland's air space is held by Norway (with the United Kingdom as a backup), which in February 2014 led a NATO training mission over Icelandic airspace in cooperation with the non-NATO countries Finland and Sweden.[86] The mission is part of a trend towards increasing cooperation between the air forces of Finland, Norway and Sweden.

## Norway

Norway views the Arctic, and to a greater extent the Norwegian Sea, as a strategic resource base that is extremely important to the national economy. It also has sovereignty issues there, especially with regard to the Svalbard Islands and the Barents Sea. However, the country has successfully resolved issues over maritime territory bilaterally with Russia. The High North, as Norway calls its part of the Arctic, is the Norwegian armed forces' most important area for strategic investment, yet it is seen as being characterized by stability.[87]

Norway seeks to increase its security in the Arctic in two ways. First, it is an active member of NATO. Norway is positive towards further engagement by the alliance in the region and is probably the member that most strongly advocates such a development. NATO is the most important institution in Norwegian defence

---

[85] The West Nordic Council is a cooperation forum of the parliaments and governments of Greenland, the Faroe Islands and Iceland. 'A parliamentary resolution on Iceland's Arctic policy' (note 83).

[86] NATO, 'Norwegian jets take on NATO's peacetime preparedness mission over Iceland', 27 Jan. 2014, <http://www.nato.int/cps/en/natolive/news_106631.htm>.

[87] Norwegian Ministry of Defence, 'Et forsvar for vår tid' [A defence for our time], Prop 73S, <https://www.regjeringen.no/globalassets/upload/fd/temadokumenter/ltp-presentasjon-fmin_komprimert_siste.pdf>, p. 30.

policy and the Washington Treaty Article V commitment to collective defence is a key component of Norwegian defence doctrine, which also extends into the High North.[88] Second, it engages bilaterally with Russia in the region. In 2010 Norway and Russia settled a 40-year dispute over maritime boundaries in the Barents Sea. Moreover, their navies exercise together biannually. There are also several initiatives along the Russian–Norwegian border, including a programme to make Norwegian visas more accessible to Russians.

In January 2015 representatives from Norway, Sweden and Finland presented a report on how to deepen Arctic cooperation between the three countries. The report suggested a number of areas where cooperation could be enhanced and was positively received by the three governments.[89]

## Sweden

Sweden does not have a particularly strong Arctic identity and before it took over chairmanship of the AC in 2011 there was little in terms of Swedish Arctic policy. Although strong in both polar and climate research, the Arctic as a political region was largely missing from the Swedish debate. As it took over the chairmanship, the Swedish Ministry for Foreign Affairs created a new Arctic policy, without a wider debate in parliament—a testament to the low political interest in the issue at the time. Ending in 2013, Sweden's chairmanship concluded a six-year period of coordinated Nordic chairmanships. Much of the work from that period came to fruition at the 2013 Kiruna ministerial meeting, including settlement of the very complicated issue of permanent observers. It was a period that strengthened the AC as an institution, with the establishment of a secretariat and with two legally binding agreements reached on SAR and oil-spill preparedness and response. Not being

---

[88] O'Dwyer, G., 'NATO rejects direct Arctic presence', *Defence News*, 29 May 2013, <http://www.defensenews.com/article/20130529/DEFREG/305290022/NATO-Rejects-Direct-Arctic-Presence>.

[89] Solberg, E., Löfven, S. and Stubb, A., 'Debatt: fortsatt samarbete med Ryssland om Arktis' [Debate: continued cooperation with Russia on the Arctic], *Dagens Industri*, 19 Jan. 2015, <http://www.di.se/di/artiklar/2015/1/20/debatt-fortsatt-samarbete-med-ryssland-om-arktis/>.

an Arctic littoral state and not being perceived as having strong national interests in the region allowed Sweden to act as an honest broker and helps to explain its diplomatic success.

Sweden's political and military leaderships continuously downplay direct military threats to the country, including from Russia. The assessment of the Swedish Defence Commission is that 'Cooperation in the Arctic is characterized by a broad consensus and a low level of conflict even though there are significant challenges'.[90] Similarly, while participating in the 2013 Arctic CHOD meeting in Greenland, the Swedish Supreme Commander, Sverker Göranson, stated that military issues were not primary but military resources were important when it came to safety issues in the region.[91] Thus, Sweden downplays the existence of 'hard' security challenges in the Arctic, but emphasizes that those challenges that do exist have a security component, albeit in a broader definition of security. While the AC is excluded from discussing matters related to 'hard' security, Swedish officials have repeatedly stated that it is useful in addressing 'soft' security challenges: 'The work of the Arctic Council gives a clear security policy value, the forum discusses in particular the broad security concept.'[92] Instead of military security, Swedish Arctic policy focuses more on issues of resilience and adaption to change.[93]

## V. An emerging circumpolar security architecture

In 1987 Soviet President Mikhail Gorbachev gave a seminal speech in Murmansk, promoting the idea of the Arctic as a zone of peace by suggesting a number of measures to increase international cooperation in the region. The Murmansk Initiative mixed soft

---

[90] Swedish Ministry of Defence, *Vägval i en globaliserad värld* [Crossroads in a globalized world], Ministry Publication Series Ds 2013:33, 31 May 2013, p. 116.

[91] Swedish Armed Forces' YouTube channel, 18 June 2013, <https://www.youtube.com/watch?v=hOUaGHZX7Zw>.

[92] Swedish Parliament, *Vägval i en Globaliserad Värld* [Choices in a Globalized World], Ministry Publication Series Ds 2013:33 (Swedish Government: Stockholm, 2013), <http://www.riksdagen.se/sv/Dokument-Lagar/Utredningar/Departementsserien/Vagval-i-en-globaliserad-varld_H1B433/>, p. 117.

[93] Swedish Ministry of Foreign Affairs, Sweden's Strategy for the Arctic Region, 2011, <http://www.regeringen.se/contentassets/2c099049a492447b81829eb3f2b8033c/sveriges-strategi-for-den-arktiska-regionen>.

issues, such as cooperation on the environment and research, with harder security issues, such as conventional arms control and a possible nuclear-free zone in the north. It signalled a shift in cold war politics towards the Arctic and, although some of the suggestions did not come into effect at the time, many of them are realities today. The Arctic may not be a nuclear-free zone, but Gorbachev's suggestions concerning the softer security issues have indirectly led to a decrease in tensions and the de-securitization of several Arctic issues.[94]

The speech also contributed greatly to region building in the Arctic by emphasizing issues beyond the previously dominant military perspective. In particular, in the areas of science and the environment, new circumpolar institutions could be developed, including the AC. Gorbachev also opened up the discussion for the idea of the eight-state region, which is the most common political conceptualization of the Arctic today. The AC has done much to create shared norms and trust in the region but what is occurring is not regionalism in the sense of, for example, the EU, which is built on integration and shared sovereignty. Rather, the successful multilateralism of the Arctic region is aimed at strengthening the sovereignty of the Arctic states. For example, the new SAR agreement that was negotiated within the AC deals more with dividing the region into sectors of responsibility than building capacity together (through e.g. pooling and sharing).[95] Further, the criteria for new permanent observers in the AC also stress the need for applicants to respect the sovereignty of the Arctic states.[96]

Over the past few years there has been a surge in the number of joint military exercises conducted in the Arctic. Canada's yearly exercise, Operation Nanook, which takes place in the Canadian Arctic, has grown to include US and Danish forces. The Norwegian-led Partnership for Peace (PFP) exercise, Cold Response, has also grown in size and 16 000 troops from 16 participating countries took part in northern Norway and Sweden in

---

[94] Åtland, K., 'Mikhail Gorbachev, the Murmansk Initiative and the desecuritization of interstate relations in the Arctic', *Cooperation and Conflict*, vol. 43, no. 3 (2008).
[95] On the SAR agreement, see <http://library.arcticportal.org/1874/1/Arctic_SAR_Agreement_EN_FINAL_for_signature_21-Apr-2011%20(1).pdf>.
[96] For a list of the criteria see the Arctic Council website, <http://www.arctic-council.org/index.php/en/about-us/arctic-council/observers>.

2014.⁹⁷ Norway also carries out a biannual joint naval exercise with Russia, Pomor, in which the two countries train and prepare for piracy and terrorism at sea, anti-submarine warfare and SAR in the Barents Sea. NORAD hosts several joint exercises in the Arctic, many of which are bilateral exercises between the USA and Canada, such as Vigilant Shield and Determined Dragon. Some also include Russian forces, for example, Vigilant Eagle that deals with terrorism in Alaskan and Russian airspace.⁹⁸ There are also bilateral exercises between the USA and Russia, for example, Northern Eagle that focuses on interoperability and SAR, but which also focused on terrorism and arms trafficking in 2004 and 2006. Northern Eagle also included Norwegian forces in 2008 and 2012.⁹⁹ Through the framework of NORDEFCO, Sweden and Finland have been exercising fighter jet tactics in the Arctic Fighter Meet since 2003. In September 2012 the eight Arctic states had their first-ever joint SAR exercise following the 2011 SAR agreement.¹⁰⁰

Bi- and multilateral military cooperation in the Arctic is fairly widespread, notably in the North Atlantic, North American and Nordic contexts. Truly circumpolar military cooperation is still in its infancy, but there are signs of an increase in such activities. Two sets of meetings stand out in this context. In April 2012 Canada hosted a meeting between the eight Arctic states' CHOD in Goose Bay, Labrador, which included Canada's General Walter Natynczyk, Russia's General Nikolai Makarov and the head of NORAD and the US Northern Command, General Charles Jacoby. Over the two-day meeting the CHOD discussed issues of civil–military relations in the north, environmental stewardship and SAR. The ambition was that it would become an annual event and a follow-up meeting was held in Greenland in 2013, but not in 2014.

---

⁹⁷ Norwegian Armed Forces, 'Cold Response', <https://forsvaret.no/en/exercise-and-operations/exercises/cold-response>.

⁹⁸ 'Russia, US, Canada holding Vigilant Eagle 2012 antiterrorist exercise' Itar-Tass, 27 Aug. 2012, <http://www.itar-tass.com/en/c154/504395.html>.

⁹⁹ Pettersen, T., 'Exercise "Northern Eagle" has started', Barents Observer, 20 Aug. 2012, <http://barentsobserver.com/en/security/exercise-northern-eagle-has-started-20-08>.

¹⁰⁰ Arctic Council, 'First live Arctic search and rescue exercise—SAREX 2012', <http://www.arctic-council.org/index.php/en/our-work2/8-news-and-events/332-first-live-arctic-search-and-rescue-exercise-sarex-2012>.

In August 2012 USEUCOM, together with the Norwegian armed forces, also set up a military level meeting in Bodö, Norway. The Arctic Security Forces Roundtable (ASFR) gathered delegates from the eight Arctic countries as well as from France, Germany, the Netherlands and the UK. The two-day meeting focused on operational communications and coordination, as well as on domain awareness. Heading the US delegation to the meeting, Major General Mark Schissler, USEUCOM's director for policy and strategy, highlighted the importance of having Russia at the table.[101] The meeting does not attract the high-level participation of, for example, the CHOD, and the Russian delegation was comprised of staff from the Russian Embassy in Norway. In September 2014 the ASFR held a second meeting in Naantali, Finland. Again, the meeting focused on safety issues such as SAR, rather than on 'hard' security questions.[102]

Despite the positive experience of Arctic security cooperation over the last five years, and despite the absence, according to many officials, of issues requiring military solutions, the Arctic is not entirely free from problems. The region remains part of the larger security relations and policies of the Arctic states. Events surrounding the Ukrainian crisis and Western countries' reactions have revealed that there are challenges to security cooperation in the Arctic which stem from issues far beyond it—as many of the Arctic states have chosen to use the region to 'punish' Russia.

Canada is a home to a significant Ukrainian migrant population, so in the wake of Russia's involvement in Ukraine during spring 2014 it has been one of Russia's staunchest critics. Along with the introduction of a number of measures, including sanctions against numerous Russian officials, Canada has chosen the Arctic arena to express its disagreement with Russian policy in Ukraine. In particular, Canadian officials abstained from participation in a meeting for the working group on black carbon held in Moscow in April

---

[101] Vandiver, J., 'Nations seek ways to cooperate on a range of Arctic issues', *Stars and Stripes*, 30 Aug. 2012, <http://www.stripes.com/news/nations-seek-ways-to-cooperate-on-a-range-of-arctic-issues-1.187454>.
[102] Kee, R., 'Arctic security forces round table: a new way to live by an old code', United States European Command, 9 Sep. 2013, <http://www.eucom.mil/media-library/blog%20post/25348/arctic-security-forces-round-table-a-new-way-to-live-by-an-old-code>.

2014.[103] Due to the events in Ukraine, Norway and the USA are also, for the time being, discontinuing military cooperation with Russia. In the Arctic, this affects the Northern Eagle trilateral exercise with USA, Norway and Russia, which was cancelled for 2014.[104]

There is a growing concern among the Arctic states regarding Russia's territorial ambitions and its readiness to use military force to achieve them.[105] Speaking in Montreal, the former US Secretary of State, Hillary Clinton, called for Canada and the USA to unite against Russia in the Arctic. According to Clinton, President Putin is trying 'to rewrite the boundaries of post-World War II Europe' and the policy might affect other countries and territories, including those in the Arctic.[106] The Norwegian Defence Minister, Ine Marie Eriksen Søreide, also claimed that 'We are in a completely new security situation where Russia shows both the ability and the will to use military means to achieve political goals'.[107]

In May 2014 the EU and the USA imposed targeted sanctions against high-ranking Russian officials.[108] Then in July 2014 the USA imposed a second round of sanctions, limiting access to the American debt market for a number of Russian banks and companies, among them Rosneft and Novatek.[109] Later the same month the EU and the USA imposed a third round of sanctions, targeting specific companies and industries.

---

[103] Pettersen, T., 'Canada skips Arctic Council meeting over Ukraine', Barents Observer, 6 Apr. 2014, <http://barentsobserver.com/en/arctic/2014/04/canada-skips-arctic-council-meeting-over-ukraine-16-04>.

[104] Nilsen, T., 'Crimea crisis puts Barents naval exercise on hold', Barents Observer, 14 Mar. 2014, <http://barentsobserver.com/en/security/2014/03/crimea-crisis-puts-barents-naval-exercise-hold-14-03>.

[105] Norris, S., 'Despite Crimea, Western–Russian cooperation in the Arctic should continue', Eurasia Outlook, 27 Mar. 2014, <http://carnegie.ru/eurasiaoutlook/?fa=55121>.

[106] Staalesen, A., 'Hillary warns against Russia in Arctic', Barents Observer, 3 Apr. 2014, <http://barentsobserver.com/en/arctic/2014/04/hillary-warns-against-russia-arctic-03-04>.

[107] Fouche, G., 'Wary of Russia, Norway urges NATO vigilance in Arctic', 20 May 2014, <http://www.reuters.com/article/2014/05/20/us-norway-defence-russia-idUSBREA4JO HE20140520>.

[108] 'Lavrov says hysterical US policy makes Russia consider appropriate response', RIA Novosti, 14 May 2014, <http://en.ria.ru/world/20140514/189825593/Lavrov-Says-Hysterical-US-Policy-Makes-Russia-Consider.html>.

[109] Schoen, D., 'New US sanctions against Russia go further than ever before', Forbes, 17 July 2014, <http://www.forbes.com/sites/dougschoen/2014/07/17/new-us-sanctions-against-russia-go-further-than-ever-before/>.

Russia's Arctic projects have been particularly affected as sanctions have banned the export to Russia of hi-tech oil equipment needed in Arctic, deep sea, and shale extraction projects.[110] Rosneft has had to suspend its cooperation with ExxonMobil in the Kara Sea and another promising Arctic project, the Yamal LNG plant, developed jointly by Novatek and Total, is struggling to get financing from European institutions.[111]

Since the development of Arctic resources is seen not only as an economic priority but also a security one, the Russian Government has to reconsider its policy of involving Western countries in Arctic projects.[112] During the Meeting of the [Russian] Security Council on state policy in the Arctic, held right after the sanctions had been launched, Putin stated that the 'dynamic and ever-changing political and socioeconomic situation in the world ... is fraught with new risks and challenges to Russia's national interests, including those in the Arctic'.[113]

The situation in Ukraine has revealed the vulnerability of the current Arctic cooperation and underlined a clear division between the Arctic states. Although they share some understanding of security challenges in the region and have achieved a certain level of cooperation, the Arctic is still represented by two camps: Russia and the rest. The external shock of the Ukrainian crisis has meant that many of the Arctic states are re-evaluating their cooperation with Russia in the region, regardless of the fact that the problems stem from events beyond its borders.

The negative reactions of Western countries towards Russia have been reflected in the Arctic, an otherwise rather non-controversial region, and are pushing Russia to hedge the risks and diversify its political and economic partners. If Russia's earlier efforts were primarily focused on building cooperation with the Arctic states

---

[110] Baker, P., Cowell, A. and Kanter, J., 'Coordinated sanctions aim at Russia's ability to tap its oil reserves', *New York Times*, 29 July 2014.

[111] Staalesen, A., 'Frenchmen keep up pressure in Yamal', Barents Observer, 19 May 2014, <http://barentsobserver.com/en/energy/2014/05/frenchmen-keep-pressure-yamal-19-05>.

[112] 'Путин пока не будет отвечать на санкции Запада' [Putin will not respond to Western sanctions for now], 30 Apr. 2014, Lenta.ru, <http://lenta.ru/news/2014/04/30/putin>.

[113] Putin, V., Speech at the Meeting of the Security Council on state policy in the Arctic, Moscow, 22 Apr. 2014, <http://eng.kremlin.ru/news/7065>.

and using Western companies to boost its Arctic projects, the future focus might be very different. Russian experts expect that the country will increase involvement with its Asian neighbours, particularly China.[114] There were attempts to work in this direction prior to the Ukraine crisis and they might be strengthened in the future. Although it is too early to call it a fully-fledged 'turn to the East' when it comes to the Arctic, the result might be a general 'cooling down' of Arctic cooperation.

Nevertheless, it would be an exaggeration to say that the Ukraine crisis has put a stop to cooperation in the Arctic. The Russian Government has not declared any change in its Arctic policy in the aftermath of the crisis and reassured that Russia is complying with the rules of UNCLOS when it comes to delimitation of the shelf.[115] Russia filed its revised application to the CLCS in August 2015.[116] The work of the Arctic institutions, primarily the AC, has also only been affected in a minor way. For example, Russia participated in the SAO meeting in Yellowknife, Canada, in 2014—seemingly without problems—in spite of strained relations between Canada and Russia.

On the bilateral level, apart from the general freeze in military-to-military cooperation, there is little evidence to suggest that the strained relations between the USA and Russia, for instance, are affecting the rest of their cooperation in the Arctic. In fact, the two countries have even utilized their good relations in the Arctic in order to address other, more pressing issues in the past. On the sidelines of the 2013 Kiruna ministerial meeting, for example, the US Secretary of State, John Kerry, and the Russian Foreign Minister, Sergei Lavrov, sat down together and discussed Syria, an issue that deeply divided the two countries at the time.[117] The uncontro-

---

[114] Trenin, D., 'Russia and China: the Russian liberals' revenge', Carnegie, 19 May 2014, <http://carnegie.ru/eurasiaoutlook/?fa=55631>.

[115] Putin, V., Meeting with heads of leading international news agencies, Transcript, 24 May 2014, <http://eng.kremlin.ru/news/7237>.

[116] Comment by the Russian Information and Press Department on Russia's application for Arctic shelf expansion, Russian Ministry of Foreign Affairs, 4 Aug. 2015, <http://www.mid.ru/web/guest/foreign_policy/news/-/asset_publisher/cKNonkJE02B w/content/id/1633205?p_p_id=101_INSTANCE_cKNonkJE02Bw&_101_INSTANCE_cK NonkJE02Bw_languageId=en_GB>.

[117] US Department of State, 'Kerry at Arctic Council Session with Swedish, Russian Ministers', Transcript, 15 May 2013, <http://iipdigital.usembassy.gov/st/english/text trans/2013/05/20130515147407.html#ixzz32uOdRHpY>.

versial nature of the Arctic may actually allow structures like the AC to help build confidence far beyond the region itself.

As previous experience and the case of Ukraine show, in the long-term perspective, further distancing between Russia and the rest of the Arctic states is very possible. Western European and North American Arctic states are strengthening ties within their frameworks of cooperation, such as the EU and NATO, while Russia is left outside those arrangements. The resulting increase in NATO and EU activities could trigger a negative reaction from Russia and a further increase in its military capabilities. In terms of political cooperation, in line with its general foreign policy directions, Russia might be drifting even further away from the West to the East. Thus, once the sovereignty of the Arctic states is ensured through the delimitation process, there might be little ground for further cooperation. Yet experience also shows a number of positive examples of the Arctic states managing to overcome their differences and developing common tools to address Arctic security issues. Hence cooperation might continue—but just not spread beyond the Arctic Circle.

## VI. Conclusions

As outlined by Alyson Bailes, there are several different approaches to Arctic security, and it is clear that states in the region view Arctic security in very different ways.[118] The Arctic states themselves can be split and grouped geographically, but other factors, such as domestic politics, influence their approaches as much as geographical determinants do. Within these approaches, a few different groupings emerge.

Canada and Russia are very focused on the sovereignty of their Arctic territories, they are not above suggesting military solutions in order to assert that sovereignty and they are not keen on opening up Arctic cooperation to outsiders. As such, Canada and Russia only reluctantly accept new permanent observers in the AC and they strongly oppose an increased NATO presence. They have vast Arctic territories, but also strong Arctic identities and the Arctic figures prominently in domestic politics.

[118] See chapter 2 in this volume.

For Iceland and Greenland's self-rule, the changing Arctic is not a question of a remote periphery, or an abstract future, but rather a practical and immediate political issue. These nations do not have military forces at their immediate disposal, and their focus is rather on practical issues such as resource management. Being small nations, they look favourably on the consensus-based AC, though Greenland would prefer to have more influence in the organization.

Norway falls somewhere in between these two groupings: it certainly has a focus on sovereignty and sees a role for the military in upholding it, but it is also very open to multilateral cooperation. Norway is positive towards new observers and looks favourably on an increased NATO presence.

Sweden and Finland do not have strong Arctic identities and their geographic location, cut off from the Arctic Ocean, gives the impression that they do not have strong national interests in the region. For this reason, they are perceived as neutral in Arctic diplomacy and are able to act as honest brokers in the multilateral cooperation of which they are strong supporters. Hard security concerns in these countries are more generally related to Russia and are not directly linked to the Arctic. Yet Sweden and Finland's strong capacities in research and ice breaking make them respected members of the Arctic community.

Denmark and the USA share the experience of having their Arctic policy subjected to a domestic dynamic between the mainland and the perceived periphery. In the USA, this relationship is between the federal government and the state of Alaska; in Denmark, it is between the government in Copenhagen and Greenland's self-rule. Foreign and security policy is formulated in Washington, DC, and Copenhagen, sometimes to the frustration of Juneau and Nuuk, and the exchange of influence and resources between mainland and periphery becomes central to the domestic debate. For both Denmark and the USA, the 'hard' security concerns are found elsewhere.

It is very possible that circumpolar security arrangements, following the examples of fully-fledged organizations such as the EU and NATO, will not come about in the Arctic. Despite some common understanding of the security concerns in the region, the Arctic states can differ greatly when it comes to addressing them.

The strong focus on sovereignty of some and the strong interconnectedness with certain domestic issues does not allow them to commit to a fully-fledged organization with 'shared' sovereignty. The Arctic states are also far from forming a genuine security community in the region, not because of differences of opinion about what Arctic security entails, but rather because of differences in their broader foreign and security policies. They clearly differ in their approaches to addressing international issues, with the crisis in Ukraine being just one example of this. The resulting revelation is that two camps still exist in the Arctic: Russia and the rest.

At the same time, this does not mean that existing security arrangements, formed in an ad hoc manner or developed within other institutions, would not succeed in addressing security issues in the region. In fact, the strong focus on sovereignty from all the Arctic states and the wish to draw up the rules of the game without too much outside interference may actually be fertile ground for extensive cooperation on security matters in the Arctic. Formalized and recurring meetings on Arctic security and an array of bi- and multilateral military exercises, against a background of mutual confidence reinforced by the AC and the 2008 Ilulissat Declaration, testify that some security concerns can be effectively addressed by the existing system. It is, however, a system that remains vulnerable to events occurring outside the Arctic.

# 4. Russia's Arctic governance policies

ANDREI ZAGORSKI

## I. Introduction

Russian policies on the governance in the marine Arctic vary depending on the particular issues or sectors of economic activities concerned.[1] The common denominator is a strong emphasis on sovereignty, sovereign rights and jurisdiction as a basis for promoting regional arrangements. In turn, those regional arrangements are seen as a manifestation of the unique rights and responsibilities of the eight member states of the Arctic Council (AC): Canada, the Kingdom of Denmark, Finland, Iceland, Norway, Russia, Sweden and the United States. While recognizing the importance of multiple global regimes as applicable in the Arctic, Russia is reluctant to admit external actors into regional governance frameworks, although it intensively engages them bilaterally in the economic development of the Russian Arctic.

This chapter applies the concept of subsidiarity in order to reconstruct the implicit logic behind the variations of Russian policy towards the marine Arctic governance with a view to better understanding and explaining them.[2] The application of the concept of subsidiarity implies that, within the area of its jurisdiction, Russia would give preference to national laws and regulations.

Regional arrangements would have a subsidiary function as far as they complement national regulations and help promote regional cooperation. Such cooperation is an expression of the unique and exclusive rights and responsibilities of Arctic as opposed to non-Arctic states. Broader international solutions engaging external actors would be desired only if a problem exceeds the jurisdictions

---

[1] This chapter primarily discusses Russian policies towards the marine Arctic since governance issues related to terrestrial parts of the region are addressed exclusively as an issue of sovereignty of coastal states.

[2] Oxford Dictionaries defines subsidiarity in politics as 'the principle that a central authority should have a subsidiary function, performing only those tasks which cannot be performed at a more local level', <http://www.oxforddictionaries.com/definition/english/subsidiarity?q=subsidiarity>.

of the Arctic states and thus cannot be fixed within a purely regional format. Such issues most obviously occur in the central Arctic Ocean beyond the exclusive economic zones (EEZs) of coastal states, but also within the EEZs.

There are a number of grey areas between the three layers of governance in the marine Arctic—national, regional and global. These areas occur when different views are expressed with regard to which level of governance is most appropriate to address a specific issue. Such differences create the potential for explicit or implicit conflict between the strong emphasis on the centrality of national sovereignty, regional exceptionalism and increasing pressures to open the circle of actors involved in Arctic governance. The search for solutions to particular issues within these areas is also affected by domestic policy debates and the peculiarities of the largely traditionalist Russian Arctic discourse.

In addition, as different areas of activity in the marine Arctic are governed by different regimes and instruments and are handled through different national, regional or global institutions, the particular mix of the governance tools from different layers, and the extent to which Russia emphasizes any of them, depends on the issues under consideration.

This chapter explores some of these combinations and related dilemmas by reviewing Russia's policies on a number of governance issues in the Arctic, such as: (*a*) the applicability of the law of the sea; (*b*) the establishment of outer limits of the continental shelf; and (*c*) navigation, international fisheries and security. It further seeks to explain, against this comprehensive background and through the prism of the concept of subsidiarity, the role attributed by Russia to regional arrangements and particularly to the AC.

## II. The law of the sea

Relevant norms of general international law as well as specifically those of the law of the sea are widely perceived in Russia as an important source for marine Arctic governance.[3] This is the case

---

[3] Savas'kov, P. V., 'Правовой режим Арктики' [The legal regime of the Arctic], ed. A. V. Zagorski, *Арктика: зона мира и сотрудничества* [The Arctic: A Zone of Peace and Cooperation] (IMEMO: Moscow, 2011), pp. 38–39 (in Russian).

for a number of international instruments, including those regarding safety of life at sea (the SOLAS Convention); protection from pollution from ships (the MARPOL Convention); international regulations for preventing collisions at sea, liability and compensation for damage in connection with the carriage of hazardous and noxious substances by sea; and the 1992 Convention on Biological Diversity.

Most Russian authors and government officials underscore the centrality of the 1982 United Nations Convention on the Law of the Sea (UNCLOS) for Arctic governance.[4] Several authors, however, emphasize the importance of international customary law based on historic practices. They suggest that Arctic customs should inform states' policies, particularly as regards the delineation and delimitation of the continental shelf or management of vessel traffic, rather than the relevant UNCLOS provisions.[5]

---

[4] In force from 1994, the Russian Federation ratified UNCLOS in 1997. United Nations Convention on the Law of the Sea (UNCLOS), <http://www.un.org/Depts/los/convention_agreements/texts/unclos/unclos_e.pdf>; Kolodkin A. L., Gutsulyak V. N. and Bobrova Yu. V., *Мировой океан. Международно-правовой режим. Основные проблемы* [The Global Ocean. International Legal Regime. Main Problems] (Statut: Moscow, 2007), pp. 259–61 (in Russian); and 'Выступление и ответы Министра иностранных дел России С.В. Лаврова на вопросы СМИ в ходе пресс-конференции по итогам переговоров с Министром иностранных дел Исландии Э. Скарпхьединссоном, Москва, 29 ноября 2011 года' [Statement and responses to media questions by Foreign Minister S. V. Lavrov at a press conference after negotiations with Foreign Minister of Iceland Ö. Skarphéðinsson, Moscow, 29 Nov. 2011], <http://www.mid.ru/bdomp/ns-reuro.nsf/348bd0da1d5a7185432569e700419c7a/c32577ca0017442b44257957004df143!OpenDocument> (in Russian).

[5] Voytolovskii, G. K., 'Нерешенные проблемы арктического морепользования' [Unresolved issues of the use of Arctic maritime areas], *Vestnik MGTU*, vol. 12, no. 1 (2010), pp. 93, 101 (in Russian); Vylegzhanin A., 'Границы континентального шельфа в Арктике: сопоставление Конвенции по морскому праву 1982 г. с обычными нормами международного права' [Limits of Continental Shelf in the Arctic: A Comparison of the 1982 Law of the Sea Convention and Customary Norms of International Law], *Международное право и национальные интересы Российской Федерации* [International Law and National Interests of the Russian Federation] (Vostok-Zapad: Moscow, 2008), pp. 26–76 (in Russian); Vylegzhanin A., 'Становление глобального правового пространства в XXI веке' [Development of a global legal space in the 21st century], *Международные процессы. Журнал теории международных отношений и мировой политики* [International Processes. A Journal of International Relations and World Politics Theory], vol. 8, no. 2 (23) (2010) (in Russian); and Vylegzhanin A. N. et al., *Предложения к дорожной карте развития международно-правовых основ сотрудничества России в Арктике* [Proposals for the Road Map of Developing International Legal Foundations of Russia's Cooperation in the Arctic], Russian International Affairs Council (Spetskniga: Moscow, 2013), pp. 28–34 (in Russian).

Despite differences among Russian experts, applicable international law is supposed to be of utmost importance for the contemporary Russian debate on Arctic governance.

First and foremost, it is considered crucial for *defining the extent of sovereignty, sovereign rights and jurisdiction*, meaning the rights and responsibilities of coastal states in the Arctic Ocean, and governing peaceful settlement of remaining and eventual disputes among them. However, it also implies the recognition of the legitimate rights and responsibilities of non-Arctic states.

Russia defines the extent of 'its' marine Arctic explicitly in terms of the law of the sea as codified in UNCLOS. According to official Russian Arctic doctrines, the territorial sea (12 nautical miles from the baselines), the EEZ (200 miles) and the continental shelf in the Arctic Ocean are included within this definition.[6] The emphasis on the relevant law of the sea provisions is instrumental for asserting the primacy of Russia's sovereign rights and jurisdiction as far as the utilization of living resources and the development of mineral resources within the EEZ and on the continental shelf are concerned.

In this context, the Russian Government particularly values consensus among the five coastal states as regards the general applicability of the law of the sea to the Arctic Ocean, as expressed in the 2008 Ilulissat Declaration by foreign ministers from the five coastal states:

By virtue of their sovereignty, sovereign rights and jurisdiction in large areas of the Arctic Ocean the five coastal states are in a unique position to address these possibilities and challenges. In this regard, we recall that an extensive international legal framework applies to the Arctic Ocean ... Notably, the law of the sea provides for important rights and obligations concerning the delineation of the outer limits of the continental shelf, the protection of the marine environment, including ice-covered areas, freedom of navigation, marine scientific research, and other uses of the

---

[6] 'Основы государственной политики Российской Федерации в Арктике на период до 2020 года и дальнейшую перспективу. Утверждена Президентом Российской Федерации 18 сентября 2008 г.' [Fundamentals of the state policy of the Russian Federation in the Arctic up to 2020 and beyond. Approved by the President of the Russian Federation on 18 Sep. 2008], section I, para. 2, <http://www.scrf.gov.ru/documents/98.html> (in Russian).

sea. We remain committed to this legal framework and to the orderly settlement of any possible overlapping claims.[7]

Some experts point out that the Ilulissat Declaration does not explicitly refer to UNCLOS but, instead, to the 'extensive international legal framework'. They imply that the declaration thus means not specifically UNCLOS but a wider framework of the law of the sea instruments (such as the 1958 Convention on the Continental Shelf) as well as customary law.

Second, the emphasis on the law of the sea and particularly on UNCLOS is instrumental in *shielding proposals to treat the Arctic Ocean as part of the global commons* and to *internationalize the Arctic* by imposing on it a global regime similar to the 1959 Antarctic Treaty and related agreements.[8] The assertion of the applicability of the law of the sea highlights the distinction between the Arctic Ocean, on the one hand, surrounded by the landmass of coastal states and the Antarctic, on the other hand, a continent surrounded by the ocean. This highlights the inappropriateness of the analogy between the Arctic and the Antarctic.

It is not surprising that the Ilulissat Declaration emphasizes that the law of the sea 'provides a solid foundation for responsible management by the five coastal States and other users of this [Arctic] Ocean'. It concludes that coastal states 'see no need to develop a new comprehensive international legal regime to govern the Arctic Ocean'.[9]

Third, the Russian Arctic governance discourse is also strongly affected by the *sectorial approach*, which is supposed to imply a legal priority of coastal states in the region within their sectors, as defined by meridian lines connecting eastern and western points of their Arctic coastlines with the North Pole.[10] However, it is widely

---

[7] 'The Ilulissat Declaration', Arctic Ocean Conference, Ilulissat, Greenland, 27–29 May 2008, <http://www.oceanlaw.org/downloads/arctic/Ilulissat_Declaration.pdf>.

[8] Proposals of this kind have been advocated by environmental groups as well as by experts, particularly from non-Arctic states (although not by the governments of those states). See, in particular, Bennet, M., 'Bounding nature: conservation and sovereignty in the Canadian and Russian Arctic', ed. L. Heininen, *Arctic Yearbook 2013* (Northern Research Forum: Akureyri, 2013), p. 86; and Lackenbauer, P. W., 'India's arctic engagement: Emerging perspectives', ed. L. Heininen, *Arctic Yearbook 2013* (Northern Research Forum: Akureyri, 2013), pp. 33–52.

[9] 'The Ilulissat Declaration' (note 7).

[10] Vylegzhanin A. N., 'Правовой режим Арктики' [Legal regime of Arctic], ed. A. N. Vylegzhanin, *Международное право*: учебник [International Law: Textbook]

acknowledged that the sectorial approach is of limited utility, as it does not affect legal regimes that apply to marine areas within the 'sectors' wherever they have been proclaimed.[11]

In sum, the recognition of the importance of the law of the sea largely serves the purpose of emphasizing sovereignty, sovereign rights and jurisdictions of coastal states as a primary level of governance. This policy is boldly expressed in the criteria for observers to the AC, adopted in 2011 at the Nuuk ministerial meeting. According to those criteria, observers are supposed to recognize 'Arctic States' sovereignty, sovereign rights and jurisdiction in the Arctic' and to recognize 'that an extensive legal framework applies to the Arctic Ocean including, notably, the Law of the Sea, and that this framework provides a solid foundation for responsible management of this ocean'.

Apart from reconfirming that the law of the sea is to be enforced 'through national implementation and application of relevant provisions', this policy aims at legitimizing the special role of the coastal states 'by virtue of their sovereignty, sovereign rights and jurisdiction in large areas of the Arctic Ocean'.[12]

## III. The continental shelf

The seabed and subsoil of submarine areas within the EEZs is considered the continental shelf of coastal states *ipso facto*. Here, coastal states exercise sovereign rights for the purpose of exploring and exploiting natural resources. They can also claim that their shelf extends beyond the EEZ. UNCLOS (Article 76, paragraph 2) provides a legal definition of the continental shelf based on a number of geomorphological criteria to prove that the respective areas beyond an EEZ are the natural prolongation of coastal states' land territory.

In order to establish the outer limits of the continental shelf, states parties to UNCLOS are entitled to submit, within ten years of its ratification, relevant evidence to the independent Commission

---

(Juright Publishers: Moscow, 2011), pp. 183–84, 204. For the discussion of the sectorial approach, see also Savas'kov (note 3), pp. 30–34.
[11] Kolodkin, Gutsulyak and Bobrova (note 4), p. 261.
[12] 'The Ilulissat Declaration' (note 7).

on the Limits of the Continental Shelf (CLCS) for review (Article 76, paragraph 8). Outer limits of the shelf established on the basis of the Commission's recommendations are considered final and binding, meaning they shall not be contested by other states parties. Where the delineated limits of extended shelf overlap, the states concerned should agree on the delimitation of their shelf before establishing its final limits (Article 83).

The area of the seabed and ocean floor beyond the limits of national jurisdiction of coastal states is considered the 'common heritage of mankind' (hereafter referred to as the 'common heritage of humanity'). The International Seabed Authority administers the exploration and exploitation of mineral resources in this area on behalf and for the benefit of humanity.

States non-parties to UNCLOS, or those who have failed to file their submission to the CLCS, cannot be denied the right to the extended continental shelf. However, they cannot benefit from UNCLOS provisions and proceed through the CLCS, with the consequence that outer limits of the shelf established by them would not be binding for other states. As a result, it 'will be for the State concerned to provide for the legal certainty necessary for exploration and exploitation activities through different means, in particular by seeking the acceptance of that claim by the State community'.[13] At the same time, the provisions related to the common heritage of humanity or the Seabed Authority do not bind states non-parties to UNCLOS. All these provisions are important for the domestic Russian debate.

Russia was the first country to present its submission to the CLCS in 2001. It was a comprehensive submission claiming several areas to which the Russian shelf would extend beyond the EEZ, including in the Arctic Ocean. In 2003 the Commission asked for more data to support this claim. It took Russian scientists more than ten years to collect data in support of a renewed submission,

---

[13] Wolfrum, R., 'The Outer Continental Shelf: Some Considerations Concerning Applications and the Potential Role of the International Tribunal for the Law of the Sea', Statement by the President of the International Tribunal for the Law of the Sea, Rio de Janeiro, 21 Aug. 2008, <https://www.itlos.org/fileadmin/itlos/documents/statements_of_president/wolfrum/ila_rio_210808_eng.pdf>, p. 7.

and a revised version was finally presented to the CLCS in August 2015.[14]

Norway is the first and so far the only Arctic state that has successfully completed the process. In 2009 the CLCS issued recommendations accepting the 2006 Norwegian claim.[15] Included in the Commission's recommendations was the need to finalize the delimitation of the maritime boundary between Norway and Russia in the grey area of the Barents Sea—a requirement that was met by the 2010 Russo-Norwegian Treaty. It is important to note that the Norwegian claim did not extend to the North Pole and stopped short of the area of the oceanic Gakkel Ridge (see figure 4.1), which does not fall under the UNCLOS definition of the continental shelf.

In December 2014, Denmark made a submission to the CLCS regarding the northern continental shelf of Greenland, indicating potential overlap with forthcoming claims by Russia, Canada and the USA, as well as recognizing the overlap with the 2006 claim of Norway in the Norwegian Sea.[16]

As Kristofer Bergh and Ekaterina Klimenko note in chapter 3, at the time of writing, Canada has not yet submitted its final claim to the extended continental shelf towards the North Pole. The USA is the only Arctic coastal state that has not yet ratified UNCLOS and is therefore not (yet) entitled to submit a claim to the CLCS. Conversely, the USA is not bound by UNCLOS and can pursue other options for extending its continental shelf, including the one based on the 1958 Convention, which neither lists specific criteria of the shelf, nor introduces any strict limit to its breadth.[17] This adds

---

[14] Partial Revised Submission of the Russian Federation to the Commission on the Limits of the Continental Shelf in Respect of the Continental Shelf of the Russian Federation in the Arctic Ocean, Executive Summary, 2015, <http://www.un.org/depts/los/clcs_new/submissions_files/rus01_rev15/2015_08_03_Exec_Summary_English.pdf>.

[15] UNCLOS Commission on the Limits of the Continental Shelf, 'Summary of the recommendations of the Commission on the Limits of the Continental Shelf in regard to the submission made by Norway in respect of areas in the Arctic Ocean, the Barents Sea and the Norwegian Sea', 27 Nov. 2006, <http://www.un.org/Depts/los/clcs_new/submissions_files/nor06/nor_rec_summ.pdf>.

[16] Partial Submission of the Government of the Kingdom of Denmark together with the Government of Greenland to the Commission on the Limits of the Continental Shelf: The Northern Continental Shelf of Greenland, Executive Summary, 15 Dec. 2014, <http://www.un.org/Depts/los/clcs_new/submissions_files/submission_dnk_76_2014.htm>.

[17] The US Government has explored different options to define the outer limits of its continental shelf, but the option to proceed on the basis of the provisions of the 1958 Convention was not considered appropriate since it does not provide sufficient legal

ambiguity to the Russian domestic debate over the limits of the continental shelf in the Arctic.[18]

The Russian Government has strictly followed UNCLOS provisions in seeking to establish extended continental shelf limits in the Arctic Ocean. Similarly to Norway, the Russian Government did not include the Gakkel Ridge area in its submissions (see figure 4.2). Furthermore, it explicitly marked it as representing the common heritage of humanity.

However, this policy is contested by large parts of the Russian expert community as well as by the political establishment (discussed further below). Inspired by the traditional sectorial approach, the critiques suggest that following the provisions of Article 76 of UNCLOS, Russia should voluntarily 'give away' part of its possessions within the sector declared in 1926—a large part of the seabed of the Arctic Ocean, which otherwise should be considered part of the Russian continental shelf.[19]

---

certainty. See US Senate Committee on Foreign Relations, 'Testimony of John B. Bellinger III, Partner, Arnold & Porter LLP', 14 June 2012, <http://www.foreign.senate.gov/imo/media/doc/John_Bellinger_Testimony.pdf>. For this reason, the US policy is based on the recognition that 'the most effective way to achieve international recognition and legal certainty for our extended continental shelf is through the procedure available to States Parties to the U.N. Convention on the Law of the Sea', see 'National Security Presidential Directive and Homeland Security Presidential Directive', 9 Jan. 2009, <http://fas.org/irp/offdocs/nspd/nspd-66.htm>. This policy is echoed by the 2013 US national Arctic strategy: 'Only by joining the Convention can we maximize legal certainty and best secure international recognition of our sovereign rights with respect to the US extended continental shelf in the Arctic', see White House, 'National Strategy for the Arctic Region', May 2013, <https://www.whitehouse.gov/sites/default/files/docs/nat_arctic_strategy.pdf>, p. 9. Research conducted by the USA in order to map the ocean floor in the Arctic and define the limits of its shelf is based on the criteria defined in Article 76 of UNCLOS; see Center for Coastal and Ocean Mapping, Joint Hydrographic Center, US UNCLOS Bathymetry Project, <http://ccom.unh.edu/theme/law-sea>. Nevertheless, as long as the USA has not ratified UNCLOS, the ambiguity of its policy persists.

[18] Vylegzhanin, in particular, asserts that the USA benefits from not being party to UNCLOS and does not need to abide by its limitations. Invoking the 1958 Convention, it can eventually extend its continental shelf almost unlimitedly through the North Pole. See Vylegzhanin (note 10), pp. 196, 199.

[19] Voytolovskii (note 5), pp. 99–102.

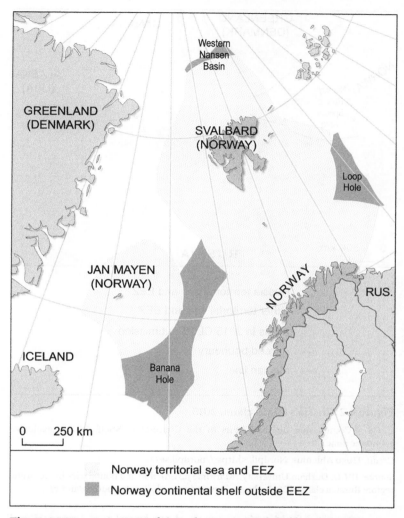

**Figure 4.1.** Norway's CLCS claims, the continental shelf outside of the EEZ

CLCS = Commission on the Limits of the Continental Shelf; EEZ = exclusive economic zone.

*Credit*: Hugo Ahlenius, Nordpil, <https://nordpil.se/>.

*Source*: Norwegian Ministry of Foreign Affairs, 'Kart over kontinentalsokkelen' [Map of the continental shelf], 2014, <https://www.regjeringen.no/globalassets/upload/ud/vedlegg/folkerett/sokkelkart_2.pdf>.

**Figure 4.2.** Russia's CLCS claim, 2015

CLCS = Commission on the Limits of the Continental Shelf; EEZ = exclusive economic zone.

*Credit*: Hugo Ahlenius, Nordpil, <https://nordpil.se/>.

*Source*: IBRU, Durham University, 'Maritime jurisdiction and boundaries in the Arctic region: Russian claims', 2015, <https://www.dur.ac.uk/ibru/resources/arctic/>.

Supporters of this view argue that, based on international customary law, the emergence of an area of the common heritage of humanity in the Arctic Ocean can be prevented in the interest of Russia and other coastal states. Instead of following the provisions of Article 76 of UNCLOS, they suggest coastal states should proceed on the basis of either the 1958 Convention, of which they all are parties, or on the basis of Article 83 of UNCLOS (delimitation of continental shelf).[20]

Although this traditionalist view of Russian possessions extending in the Arctic through the North Pole does not prevail over the government's policy, it was boldly articulated at a meeting of the advisory board of the Russian Maritime Collegium in April 2008.[21] Ever since, it has widely informed opinion within the collegium and in both chambers of the parliament. Traces of this approach can be identified in the official Russian Arctic policy doctrines. In particular, the 2013 Strategy emphasizes that, while establishing outer limits of the Russian continental shelf in the Arctic Ocean, the government shall 'prevent any losses' and ensure that legal conditions for Russian operations in the Arctic are indiscriminate as compared to other coastal states (notably the USA).[22]

This internal debate produces a certain ambiguity concerning the future course of Russian policy in establishing outer limits of the continental shelf in the Arctic, particularly if the Russian Government fails to substantiate its claim as submitted in 2015. Will Russia accept the recommendations of the CLCS or seek to assert its sovereign rights on the continental shelf outside the UNCLOS framework by referring to customary law as well as to the 1958 Convention, as the traditionalists suggest? The ratification of UNCLOS by the USA would be an important argument to help keep Russia's policy within the Convention's cooperative framework, as it would deprive the traditionalists in Russia of one of their strongest arguments.

## IV. Navigation

In accordance with the law of the sea, all states enjoy freedom of navigation and overflight in EEZs beyond the limits of the territorial sea (Article 58 of UNCLOS). However, Canada and Russia

---

[20] Voytolovskii (note 5), pp. 93, 100–101; and Vylegzhanin (note 10), pp. 196–97.
[21] Voytolovskii (note 5), p. 99.
[22] 'Стратегия развития Арктической зоны Российской Федерации и обеспечения национальной безопасности на период до 2020 года. Утверждена Президентом Российской Федерации 20 февраля 2013' [Strategy for developing the Arctic Zone of the Russian Federation and providing national security until 2020. Approved by the President of the Russian Federation on 20 Feb. 2013], <http://www.consultant.ru/document/cons_doc_LAW_142561/>, para. 29 (in Russian).

practice extensive regulation of vessel traffic in Arctic waters. These practices found confirmation in Article 234 of UNCLOS, which places the environmental jurisdiction of coastal states in ice-covered waters over the general freedom of navigation:

Coastal States have the right to adopt and enforce non-discriminatory laws and regulations for the prevention, reduction and control of marine pollution from vessels in ice-covered areas within the limits of the executive economic zone, where particularly severe climatic conditions and the presence of ice covering such areas for most of the year create obstructions or exceptional hazards to navigation, and pollution of the marine environment could cause major harm to or irreversible disturbance of the ecological balance.[23]

To take advantage of this clause, coastal states do not need to submit such regulations for endorsement by the International Maritime Organization (IMO).

### The Northern Sea Route

During the cold war the Soviet Union sought to assert its jurisdiction over the Northern Sea Route (NSR).[24] This was done primarily for national security reasons, taking into account the importance of the Russian North for the operation of strategic forces, as well as regular US submarine activities in the area.[25] For decades the NSR was closed to foreign vessels and began opening only towards the end of the cold war. After the end of the cold war Russia continued to regulate navigation through the NSR, and even expanded the area to which such regulations applied.

In 1964 Russia unsuccessfully claimed the Dmitriy Laptev and Sannikov straights connecting the Laptev and the Eastern Siberian Seas (which were too broad to fall under the territorial sea definition) to be Soviet (Russian) historic waters. It also claimed that

---

[23] UNCLOS (note 4).
[24] The NSR includes water areas adjacent to the Russian Arctic coastline, except for the Barents and Pechora seas. In the West it begins with the Kara Gate, a strait connecting the Pechora and Kara seas.
[25] Arbatov, A. and Dvorkin, V., 'Military-strategic activity of Russia and the U.S.', eds A. A. Dynkin and N. I. Ivanova, *Russia in a Polycentric World* (Ves Mir: Moscow, 2012), pp. 473–80.

navigation through those straights, as through the entire NSR, was governed by Soviet laws applying to territorial and internal sea waters.[26]

In 1984 the Soviet Union adjusted its claim to exercise jurisdiction in the NSR, building on the language of Article 234 of UNCLOS. In 1984 a decree tasked the government with establishing special rules of navigation in the NSR for the purpose of strengthening environmental protection in marine areas adjoining the northern coast of the Soviet Union. Such rules were subsequently adopted in 1990.[27] The 1998 Russian Law on internal sea waters, territorial sea and contiguous zone declared the NSR 'a historic national transport line of the Russian Federation', de facto equalling its routes to internal waters.[28]

A 'Northern Sea Route Law' adopted in July 2012 amended several previous navigation instruments in the NSR.[29] It reconfirmed the NSR status as 'a historic national transport line of the Russian Federation', re-established the NSR Administration—an agency under the Russian Ministry of Transport that is tasked with administering the rules of navigation in the NSR—and, most notably, expanded the area of Russia's environmental jurisdiction to include the entire EEZ. The revised rules of the NSR navigation were approved in January 2013.[30] Following previous practices, receiving a permit for a vessel to sail in the NSR is conditional on meeting specific Arctic class requirements as regards design, construction and equipment of ships, as well as crew qualifications. Sailing in Russia's Arctic waters may require icebreaker escort and/or the assistance of an ice pilot. Particular requirements depend on the ice situation at the time of navigation in a specific part of the area.

---

[26] Savas'kov (note 3), p. 32.
[27] Savas'kov (note 3), p. 37.
[28] Savas'kov (note 3), p. 32.
[29] The Northern Sea Route Administration, 'The Federal Law of July 28, 2012, No 132-FZ on amendments to certain legislative acts of the Russian Federation concerning state regulation of merchant shipping in the area of the Northern Sea Route', <http://www.nsra.ru/ru/zakon_o_smp/>. See also *Arctic Herald*, no. 3 (2012), pp. 68–72.
[30] 'Rules of navigation in the water area of the Northern Sea Route', Approved by the order of the Ministry of Transport of Russia, 17 Jan. 2013, <http://www.nsra.ru/files/fileslist/20150513153104en-Rules_Perevod_CNIIMF-13%2005%202015.pdf>.

The conceptualization of the NSR as Russia's historic waters is deeply rooted in the Russian establishment and society. Some authors even speak of Russia's sovereignty over the NSR.[31] Others go as far as to suggest that coastal states should extend their environmental jurisdiction beyond EEZs to cover their entire marine Arctic 'sectors'.[32]

This discourse is reinforced by traditional national security arguments, as well as by the fact that Russian policies of asserting jurisdiction over the NSR have been continuously challenged by other nations, most notably by the USA. Although generally accepting the provisions of Article 234 of UNCLOS, the USA interprets them in the context of Article 236 (sovereign immunity), which stipulates that the 'provisions of this Convention regarding the protection and preservation of the marine environment do not apply to any warship, naval auxiliary, other vessels or aircraft owned or operated by a State and used, for the time being, only on government non-commercial service'.[33]

The domestic Russian debate over the legal grounds for extensive national regulation of vessel traffic through the NSR has in recent years been fuelled by projections of increased seasonal navigation enabled by receding summer ice. Indeed, there has been a steady growth in traffic by a factor of 2.5 between the low point in 1998 and today. Concerns related to possible 'illegal' penetration of the NSR by foreign vessels have grown further in light of the international debate over the prospects for increased international navigation in the Arctic. The budding, although cautious, interest of non-Arctic states in the commercial shipping opportunities opening in the Arctic (discussed by Linda Jakobson and Seong-Hyon Lee in chapter 5) is seen as a challenge to be met by an improved capability to enforce rules extended to the NSR.

Projections of increased vessel traffic through the NSR, including international transit traffic, were at the heart of the revision of the relevant Russian legislation in 2012 and the restoration of the NSR Administration. Debates over the new law revived old divisions

---

[31] Voytolovskii (note 5), p. 98.
[32] Vylegzhanin (note 10), pp. 183–84.
[33] Kraska, J., 'The Law of the Sea Convention and the Northwest Passage', *International Journal of Maritime and Coastal Law*, vol. 22, no. 2 (2007), p. 274.

between the relevant government agencies and the traditionalists in both chambers of the parliament that were centred, among other things, around the rationale of reasserting the status of the NSR as a traditional national route of transportation—a formula strongly opposed by the government but no less strongly, and successfully, pursued by the parliament. This debate was the main reason for repeated delays in passing the 2012 bill.

The debate resulted in a comprehensive solution based on a fragile consensus within the Russian establishment and that emphasizes the importance of Article 234 of UNCLOS. The 2012 'Northern Sea Route Law' explicitly uses UNCLOS language, defining the NSR as an area extending to internal sea waters, territorial sea, contiguous zone and the Russian EEZ.[34] In a commentary to the law, the Russian Ministry of Transport explicitly refers to Article 234 as the legal grounds for the regulation of vessel traffic in NSR.[35] Like Canada, Russia does not extend its environmental jurisdiction in the marine Arctic beyond the 200 nautical miles limit allowed by UNCLOS.

Russian traditionalists accept this line of argument and admit the importance of Article 234 as an additional argument for Russia to exercise jurisdiction with regard to the NSR—as long as it applies to the NSR, i.e. as long as the Russian EEZ remains ice-covered for most of the year.[36] The debate over the legal grounds of the national regulation of vessel traffic through the NSR is also tempered contemporarily by a relatively modest volume of traffic (which so far remains slightly above half of the maximum volume of 1987), as well as by the fact that the volume is growing and is expected to continue to do so—primarily from destination shipping and much less from international transit traffic, which is still in its

---

[34] 'Article 3 of the Law', *Arctic Herald*, no. 3 (2012), p. 69.

[35] Klyuev, V., 'On amendments to certain legislative acts of the Russian Federation concerning state regulation of merchant shipping in the area of the Northern Sea Route', *Arctic Herald*, no. 3 (2012), p. 75. See also Выступление Представителя России в Комитете старших должностных лиц Арктического совета, Посла по особым поручениям А.В.Васильева на Международном Арктическом Форуме «Арктика территория диалога», Москва, 22–23 сентября 2010 года [Statement by the Representative of Russia in the Committee of Senior Arctic Officials, Ambassador at large A. V. Vassiliev at the International Arctic Forum 'The Arctic: Territory of Dialogue', Moscow, 22–23 Sep. 2010], <http://www.mid.ru/bdomp/ns-dos.nsf/45682f63b9f5b253432569e7004278c8/c85bcbec54d02d89c32575bc00243e13!OpenDocument> (in Russian).

[36] Voytolovskii (note 5), p. 98.

experimental phase. Apart from more appropriate weather and ice conditions, a substantial increase in vessel traffic through the NSR would require significant improvements to the route's infrastructure (ranging from improved hydrography and cartography through ensuring proper communications, to construction of deep sea ports and docking facilities, as well as significantly expanding capabilities available for area awareness and emergency relief to provide for appropriate maritime security). All Russian strategies for developing the Arctic spell out ambitious plans for improving navigation infrastructure along the NSR, but the ability of Russia to appropriately fund their implementation is relatively low.

Nevertheless, the contemporary domestic consensus on international vessel traffic in the Russian Arctic cannot be taken for granted forever. It may be challenged by various factors. If and when the Arctic ice melts down to the extent that the NSR becomes ice free for most of the year, and particularly if, against this background, growing military and commercial shipping starts challenging the NSR status, the debate concerning the legality of NSR regulations and the freedom of navigation within the Russian EEZ would reopen.

### The Polar Code

No special rules apply to navigation in the central Arctic Ocean beyond the EEZs. Instead, only general norms apply there, such as safety of life at sea (the SOLAS Convention) and protection from pollution from ships (the MARPOL Convention). While vessel traffic is expected to grow, this is increasingly recognized as a gap in Arctic governance. This area is located beyond the limits within which coastal states can claim environmental jurisdiction under Article 234 of UNCLOS. Although a regional arrangement of members of the AC was not entirely ruled out, it remained highly questionable whether an exclusive regional regime could be established in the high seas of the Arctic Ocean where all states enjoy the freedom of navigation. This is a reason why the preferred option was to develop global rules within the International Maritime Organization (IMO).³⁷

IMO started addressing the issue in 1992, with a view to establishing legally binding requirements for ships operating in ice-covered Arctic waters. Ten years later it adopted voluntary guidelines for the area.³⁸ In 2009 these guidelines were extended to ships operating in all Polar waters, including the Antarctic.³⁹ In the same year, Denmark, Norway and the USA initiated discussions on developing a mandatory International Code of Safety for Ships Operating in Polar Waters (known as the Polar Code). Since 2010 it has been subject to deliberations in the IMO.⁴⁰ In November 2014 the IMO Maritime Safety Committee approved Polar Code-related amendments to the SOLAS Convention and in May 2015 the IMO Marine Environment Protection Committee adopted environmental provisions of the Polar Code, the required amendments to the MARPOL Convention and its relevant annexes—thus finalizing the work on the Polar Code.⁴¹ Its provisions will become mandatory through amendments to the SOLAS and MARPOL conventions, to be introduced through the tacit acceptance procedures, and are expected to enter into force in January 2017.⁴²

Russia has committed to developing a mandatory polar code on a number of occasions, among other things, by endorsing the 2008 Ilulissat Declaration and the 2013 Arctic Council Kiruna Declaration.⁴³ It also engaged actively in drafting the Polar Code. However, except for a few stakeholders among the large shipping

---

³⁷ Zagorski, A. V., 'Vessel traffic regulation', A. V. Zagorski (ed.), A. I. Glubokov and E. N. Khmelyova, *International Cooperation in the Arctic. 2013 Report* (RIAC, Spetskniga: Moscow, 2013), pp. 36–37.

³⁸ International Maritime Organization (IMO), *Guidelines for Ships Operating in Polar Waters* (IMO: London, 2010).

³⁹ IMO Assembly, 26th session, Resolution A.1024(26), Adopted on 2 Dec. 2009, <https://www.imo.org/blast/blastDataHelper.asp?data_id=29985&filename=A1024(26).pdf>.

⁴⁰ Zagorski (note 37), pp. 32–37.

⁴¹ 'International Code For Ships Operating in Polar Waters (Polar Code)', Resolution MEPC.264(68) (adopted on 15 May 2015), *Report of the Marine Environment Protection Committee on its sixty-eighth session*, MEPC 68/21/Add.1, 5 June 2015, Annex 10.

⁴² International Maritime Organization (IMO), 'Shipping in polar waters. Development of an international code of safety for ships operating in polar waters (Polar Code)', <http://www.imo.org/MediaCentre/HotTopics/polar/Pages/default.aspx>.

⁴³ 'Kiruna Declaration on the occasion of the Eighth Ministerial Meeting of the Arctic Council', 15 May 2013, <https://oaarchive.arctic-council.org/bitstream/handle/11374/93/MM08_Final_Kiruna_declaration_w_signature.pdf?sequence=1&isAllowed=y>, p. 4.

companies, there were few advocates of the Code in Russia, especially since Moscow believed that Russia would be better served by national regulation.

There were several concerns voiced in Russia with regard to the Polar Code. While some were of an economic nature and go beyond the scope of this chapter, there were also concerns related to Arctic governance. In particular, the establishment of mandatory global rules created the potential for conflict between Russia's determination to enforce its jurisdiction over the NSR and the recognition of the need for improved maritime safety throughout the Arctic Ocean.

The Polar Code does not differentiate between maritime areas, such as high seas and EEZ. Its rules apply throughout the Arctic Ocean (as well as in the Antarctic), except for the territorial sea and internal waters of coastal states. On the one hand, this eventually enables Russia to extend some of its national rules to the central part of the Arctic Ocean. On the other hand, concerns were raised that the establishment of binding global rules would challenge rather than help reinforce rules exercised by Russia within the EEZ. Although the Polar Code is not supposed to alter legal regimes covering specific maritime areas in the Arctic, it remains open as to whether coastal states will be able to preserve the right, under Article 234, to maintain national rules of navigation that go beyond the provisions of the Code. This is an important question from various perspectives.

First, the Polar Code does not govern all vessel traffic in the area. It will apply only to commercial or passenger ships on international voyage under the SOLAS Convention. Although it is anticipated to establish specific rules for fishing vessels and leisure ships at a later stage, this is currently only a vague prospect. Second, the Polar Code, according to Article 236 of UNCLOS, does not cover the operation of military ships and vessels in the service of governments. Hence, were it to replace national rules, it would substantially reduce the scope of regulation of navigation in the NSR, and particularly in the areas important to Russia from a national security perspective.

Another significant issue is the enforcement of the Polar Code provisions. Will the sailing permits and ice certificates issued by relevant authorities of the flag state and/or by respective classification societies suffice for sailing in Arctic waters, or will coastal states be authorized to enforce the Code's provisions?[44]

Debates over these controversial issues reveal a pattern of Russian policy with regard to Arctic shipping governance that fully corresponds to the concept of subsidiarity. A clear preference is given to the exercise of national jurisdiction. At the same time, Moscow was open to considering a regional or a global arrangement that would apply to areas beyond its jurisdiction, but only if such an arrangement would complement and not question its jurisdiction.

## V. International fisheries

While national fisheries jurisdiction extends throughout the EEZ of a coastal state, international fisheries beyond that limit are governed by the relevant UNCLOS provisions and, more specifically, by the 1995 Agreement for the implementation of the provisions of UNCLOS relating to the conservation and management of straddling fish stocks and highly migratory fish stocks (also known as the Fish Stock Agreement).[45] Practical work to ensure conservation and rational utilization of marine biological resources, based on continuous fish stocks research, is done by regional fisheries management organizations (RFMOs) established in different parts of the global ocean.

Russia is part of several arrangements operating in the Arctic, such as the North East Atlantic Fisheries Commission (NEAFC) or the North Atlantic Salmon Conservation Organization (NASCO).[46] Research supporting NEAFC decisions is largely conducted under the auspices of the International Council for the Exploration of the Sea (ICES). Russia also participates in a number of specific

---

[44] Zagorski (note 37), p. 35; and Zagorski, A. V. et al., *The Arctic: Proposals for the International Cooperation Roadmap* (RIAC, Spetskniga: Moscow, 2012), p. 23–24.

[45] United Nations, *Agreement for the Implementation of the Provisions of the United Nations Convention on the Law of the Sea of 10 December Relating to the Conservation and Management of Straddling Fish Stocks and Highly Migratory Fish Stocks*, A/Conf. 164/37 (1995), UN Conference on Straddling Fish Stocks and Highly Migratory Fish Stocks, 24 July–4 Aug. 1995.

[46] Glubokov, A. I. and Glubokovsky, M. K., 'International fisheries governance', Dynkin and Ivanova (note 25), pp. 490–98; and Krayniy, A., 'Arctic fisheries: The present and the future', *Arctic Herald*, no. 3 (2013), pp. 20–37.

fisheries regimes operating under the auspices of NEAFC, such as coastal state conferences for herring or blue whiting. In general, Russia is satisfied with the way these RFMOs operate.

However, regional arrangements, except for NASCO, do not extend to the central Arctic Ocean, which remains an area of unregulated eventual international fishing (see figure 4.3). The reason why no RFMO has been established for this area is obvious. So far, there has been no commercial fishing in the ice-covered central Arctic Ocean. However, as ice continuously recedes in the summer, new potential fishing grounds may open in this area. Should this happen, new fishing grounds are most likely to open in the area of the Chukchi Plateau adjacent to the Russian and US EEZs.

Since 2010 the conservation of fish stocks in the central Arctic Ocean has been subject to bilateral and multilateral consultations by experts and representatives of the five Arctic coastal states. The purpose is to draft an agreement introducing a moratorium on commercial fishing in the central Arctic Ocean, until such time as an RFMO is established for the area based on solid scientific knowledge.[47] Since this area lies beyond the remit of national fisheries jurisdictions of coastal states, it is anticipated that other states, which 'may have an interest in this topic', would join the process leading to a binding international agreement.[48] As discussed by Jakobson and Lee in chapter 5, China may be one of those countries. However, the possibility of new fishing grounds opening in the Arctic is also being discussed in Japan, the Republic of Korea and India.[49]

---

[47] Vylegzhanin, A. N. et al., *International Cooperation in Environment Protection, Preservation, and Rational Management of Biological Resources in the Arctic Ocean*, Working paper from the international scientific symposium held in Moscow on 4 Sep. 2012 (RIAC, Spetskniga: Moscow, 2013), pp. 36–50, 59–68; Zagorski et al. (note 44), pp. 17–19; and Zagorski A. V., 'Agreement concerning fisheries in the Central Arctic Ocean', Zagorski (ed.), Glubokov and Khmelyova (note 37), pp. 20–26.

[48] 'Chairman's statement', Meeting on Arctic Fisheries, Nuuk, Greenland, 24–26 Feb. 2014, <http://www.pewtrusts.org/-/media/assets/2014/09/arcticnationsagreetoworkoninternationalfisheries-accord.pdf?la=en>.

[49] Highleyman, S., 'Protecting fisheries in the Central Arctic Ocean', Presentation at the international conference 'The Arctic: Region of cooperation and development', Moscow, 2–3 Dec. 2013; and Lackenbauer (note 8), pp. 39, 46.

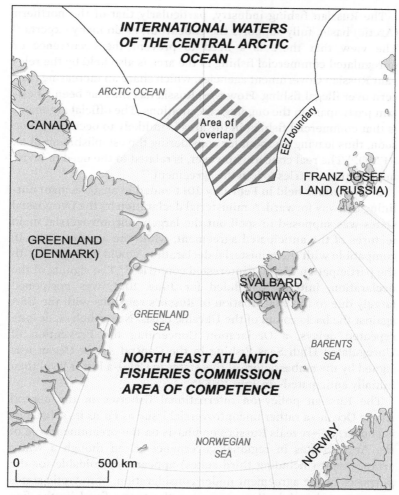

**Figure 4.3.** NEAFC and international waters in the Arctic Ocean overlap

NEAFC = North East Atlantic Fisheries Commission.

*Credit*: Hugo Ahlenius, Nordpil, <https://nordpil.se/>.

*Source*: PEW Environment Group, Map VLIZ.2014, 'World exclusive economic zones boundaries v8', 28 Feb. 2014, <http://www.marineregions.org/downloads.php>.

The Russian fishing industry, particularly that of the northern (Arctic) basin, fully supports the endeavour, as do many experts.[50] The view that there is a need to prevent the occurrence of unregulated commercial fishing in the area is also held by the relevant Russian Government agencies, which share an increasing concern over illegal fishing. However, Russia has thus far been a hesitant participant in the ongoing consultations. The official argument is that commercial fishing in the area is unlikely to occur any time soon, thus leaving no room for considering the establishment of an RFMO.[51] The real concern, however, is related to the need to invite non-Arctic countries to join the agreement.[52]

Consultations held in February 2014 ended in an agreement outlining the way forward. A ministerial declaration by the five coastal states was supposed to spell out the largely uncontroversial main features of the anticipated agreement, while the agreement itself, compatible with the ministerial declaration, would be drafted with the participation of other interested countries.[53] The signing of the declaration, initially scheduled for June 2014, was postponed largely due to the exacerbation of Russia's relations with the USA against the background of the Ukraine crisis. Nevertheless, despite repeated delays, a Declaration Concerning the Prevention of Unregulated High Seas Fishing in the Central Arctic Ocean was signed by the ambassadors of five states in Oslo—a lower level than initially anticipated—in July 2015.[54]

The Russian policy on international fisheries in the central Arctic Ocean—a rather uncontroversial issue as far as its substance is concerned—reveals Russia's emphasis on the preeminent role of the Arctic states in regional governance. Even though a wider arrangement (including third states) appears unavoidable for the purposes of the agreement under consideration, non-Arctic states are supposed to join the process on the terms fixed by the five coastal states.

---

[50] Zilanov, V. K., 'Arctic fisheries: New challenges', Vylegzhanin et al. (note 47), pp. 48–50.
[51] Krayniy (note 46), pp. 33–37.
[52] Zagorski (note 47), pp. 20–26.
[53] 'Chairman's statement' (note 48).
[54] Declaration Concerning the Prevention of Unregulated High Seas Fishing in the Central Arctic Ocean, Oslo, 16 July 2015, <https://www.regjeringen.no/globalassets/departementene/ud/vedlegg/folkerett/declaration-on-arctic-fisheries-16-july-2015.pdf>.

## VI. Security

Security is one of the least regulated areas of the marine Arctic. Rules applied here include a general ban on deploying nuclear weapons on the seabed or militarizing the ocean floor in the area of the heritage of humanity. However, these rules are yet to be delineated in the Arctic. Moreover, defence policies remain a sovereign discretion of individual states, bearing in mind that four of the five coastal states, as well as five of the eight members of the AC, are members of the North Atlantic Treaty Organization (NATO).

### Fragmented security architecture

Military deployments in the region have increased, but this is not yet the start of an arms race.[55] Although further increases in military and constabulary capabilities are anticipated (for the purpose of meeting challenges to human and environmental security), there is a growing understanding that 'the increased military attention in the High North may therefore, at least in part, be a securitization of the region rather than a militarization'.[56]

Against the background of financial austerity pressures, most Arctic states are seeking to respond to security challenges in the region by fostering regional partnership relations, rather than by substantially increasing investment in security infrastructure. However, as Alyson Bailes points out in chapter 2, 'Arctic security capabilities, activities, and relationships are fragmented between

---

[55] Wezeman, S. T., *Military capabilities in the Arctic*, SIPRI Background Paper (SIPRI: Stockholm, March 2012); Le Mière, C. and Mazo, J., *Arctic Opening: Insecurity and Opportunity* (IISS and Routledge: Abingdon and New York, 2013), pp. 77–94; Oznobishchev, S. K., 'Military activity of the Arctic States', eds Dynkin and Ivanova (note 25) pp. 465–72; Zagorski, A., 'Developing the Arctic: Security issues', eds A. Arbatov and A. Kaliadine, *Russia: Arms Control, Disarmament and International Security* (IMEMO: Moscow, 2012) pp. 116–33; Oznobishchec S. K., 'Конвенциональные вопросы безопасности в Арктике' [Conventional security issues in the Arctic], ed. Zagorski (note 3), pp. 87–102; and Zagorski A. V., 'Военная безопасность в Арктике' [Military security in the Arctic], ed. I. S. Ivanov, *Арктический регион: Проблемы международного сотрудничества: Хрестоматия в 3 томах* [*The Arctic Region: Problems of International Cooperation: A Compendium of Publications in three Volumes*] (RIAC and Aspekt Press: Moscow, 2013), vol. 1, pp. 256–69.

[56] Le Mière and Mazo (note 55), p. 95.

different national jurisdictions, bilateral relationships and groupings'. Within NATO, for example, closer security cooperation between the USA and Canada has developed over decades. More recently, among the Nordic countries, Finland and Sweden have cooperated on security issues through the NATO Partnership for Peace framework.

During the cold war, Russia was excluded from security cooperation in the Arctic. This gradually changed in the late 1990s and particularly in the 2000s, when regular trilateral naval exercises involving Russia, the USA and Norway were arranged, as well as bilateral exercises involving Russia and Norway.[57] Consequently, cooperation between those countries' coastguards and rescue agencies also intensified. Expectations of expanded cooperation between the constabulary forces—both bilaterally and multilaterally—were also raised following the 2011 Agreement on Cooperation on Aeronautical and Maritime Search and Rescue in the Arctic (SAR Agreement) by eight AC member states.[58] These were further enhanced by the 2013 Agreement on Cooperation on Marine Oil Pollution Preparedness and Response in the Arctic (Oil Spill Agreement).[59]

Overall, however, attempts to engage Russia in cooperative security frameworks in the Arctic have made limited progress, and the efforts were put on hold in 2014 while still in their infancy. The USA and Norway suspended military cooperation with Russia over the Ukraine crisis.[60] Senior level meetings were postponed, including not only those involving chiefs of defence but also

---

[57] Le Mière and Mazo (note 55), p. 96.

[58] Agreement on Cooperation on Aeronautical and Maritime Search and Rescue in the Arctic (2011), <https://oaarchive.arctic-council.org/bitstream/handle/11374/531/Arctic_SAR_Agreement_EN_FINAL_for_signature_21-Apr-2011%20%281%29.pdf?sequence=1&isAllowed=y>.

[59] Agreement on Cooperation on Marine Oil Pollution Preparedness and Response in the Arctic, 15 May 2013, <https://oaarchive.arctic-council.org/bitstream/handle/11374/529/MM08_agreement_on_oil_pollution_preparedness_and_response_%20in_the_arctic_formatted%20%282%29.pdf?sequence=1&isAllowed=y>.

[60] LaGrone, S. and Majumdar, D., 'Planning for joint U.S. and Russia naval exercise on hold pending outcome in Crimea', USNI News, 24 Mar. 2014, <http://news.usni.org/2014/03/04/planning-joint-u-s-russia-naval-exercise-hold-pending-outcome-crimea>; and Norwegian Ministry of Defence, 'Norway suspends all planned military activities with Russia', Press Release no. 25/2014, 25 Mar. 2014, <http://www.regjeringen.no/en/dep/fd/press-centre/Press-releases/20141/Norway-suspends-all-planned-military-activities-with-Russia-.html?id=753887>.

coastguard officials. The creation of a platform that would enable all the Arctic states, including Russia, to foster partnerships and discuss security-related concerns now seems a distant prospect.[61] As does any closing or narrowing of the gap in the regional security architecture.

Several options to address this issue have been proposed over the years. The first implies the possibility of empowering the AC to perform the platform function and to include a broad variety of security issues on its agenda. Choosing this route would require amending the AC's mandate, since the founding 1996 Ottawa Declaration explicitly exempted 'matters related to military security'.[62]

The second option acknowledges the fact that the majority of Arctic states are members of NATO, and suggests that the NATO–Russia Council could serve as a forum for discussing security issues and cooperation in the Arctic.

The third option builds on the emerging new cooperative frameworks, such as the annual meetings of chiefs of defence held in 2012 and 2013 (postponed in 2014) focusing on cooperative implementation of the 2011 SAR Agreement. The agenda included expanded cooperation on maritime surveillance and joint exercises, and the USA sponsored annual Arctic Security Forces Roundtable meetings attended by the eight AC member states, as well as France, Germany, the Netherlands and the UK.

Building on the cooperative experiences within the North Atlantic and the North Pacific Coast Guard Forums, and as preparation for its chairmanship of the AC (2015–17), the US Government established an Arctic Coast Guard Forum, with a view to facilitating the implementation of the SAR Agreement.[63] The first executive meeting of the forum was supposed to be organized in early 2015 and the USA is meant to take over the chairmanship of the forum during its chairmanship of the AC.

---

[61] Le Mière and Mazo (note 55), pp. 97–99; and Berkman, P. A., 'Preventing an Arctic cold war', *New York Times*, 12 Mar. 2013.

[62] Declaration on the Establishment of the Arctic Council, Ottawa, 19 Sep. 1996, <https://oaarchive.arctic-council.org/bitstream/handle/11374/85/00_ottawa_decl_1996_signed%20%284%29.pdf?sequence=1&isAllowed=y>.

[63] White House, 'Implementation Plan for The National Strategy for the Arctic Region', Jan. 2014, <https://www.whitehouse.gov/sites/default/files/docs/implementation_plan_for_the_national_strategy_for_the_arctic_region_-_fi....pdf>, p. 25.

## Russian policies

Asserting national sovereignty and sovereign rights in the marine Arctic through restoring, maintaining and developing sufficient defence and security capabilities is paramount within Russian policy. However, this policy is complemented by efforts to promote and expand regional cooperation, both bilateral and multilateral, in order to transcend contemporary fragmentation of the security architecture, promote confidence and alleviate any tensions resulting from the securitization of the region. The policy generally fits into the broader trend among Arctic nations seeking to foster regional partnerships as a more cost-efficient way of addressing new challenges in the marine Arctic.

Russian defence and security-related programmes in the Arctic are discussed in this volume by Bailes (chapter 2) and at greater length by Bergh and Klimenko (chapter 3). In the context of this chapter, it is important to note that Russia's current investment in border and coastguard capabilities (including integrated rescue centres and rescue capabilities of the Ministry of Transport and local authorities, defence infrastructure and area awareness, and the construction of three new nuclear-powered icebreakers) fits into the broader context of Russian investment in the infrastructure of the Russian Arctic in order to support growing economic activities.

As part of this strategy, restored defence facilities (air and naval) are supposed to be 'dual use', meaning they should be available for civil authorities or constabulary forces, among others, for further exploration of the marine Arctic or the provision of disaster relief. Conversely, defence infrastructure serves strategic purposes, such as improving airspace defence and anti-submarine capabilities, as well as re-establishing springboard airfields for strategic aircraft, which resumed patrolling the Arctic space in 2007. In this context, it is obvious that any proposals to declare the Arctic a nuclear-free zone are taboo in Russia, the reason being that its Northern Fleet operating in the Arctic remains a crucial component of the Russian strategic nuclear forces.

Seeking to complement the country's own defence-related activities by increased regional cooperation, not least in order to avoid unnecessary securitization effects of the build up of its own defence capabilities, Russia is flexible as regards options for pursuing cooperation, although not all options described above meet Moscow's criteria.[64]

It is particularly the option of engaging NATO or the NATO–Russia Council as a platform for security cooperation in the Arctic that is out of bounds for the Russian establishment.[65] Although Russian representatives have participated in the US-sponsored annual Arctic Security Forces Roundtable, they did not display much enthusiasm during these meetings. One reason for this was the extended invite to non-Arctic NATO member states. Indeed, this roundtable framework is barely mentioned in official Russian sources in the context of Arctic security cooperation.

Russia has a flexible approach to other options. It does not, in principle, oppose the idea of amending the mandate of the AC in order to empower it to discuss security issues. However, Russia does not pursue this option either, being aware of hesitations among some other member states, particularly the USA.

In recent years, Russia has emphasized the important role to be played by annual meetings of chiefs of defence.[66] However, the 2014 meeting was suspended as a consequence of the Ukraine crisis, and it is still unclear whether those meetings will resume any time soon.

Russia also underlines the need to develop practical cooperation within the framework of the 2011 SAR Agreement, as well as the 2013 Oil Spill Agreement. The 2013 Russian Arctic strategy anticipates the establishment of an integrated regional SAR capability as well as a capability tailored to respond to man-made disasters, based on close coordination between relevant agencies of coastal

---

[64] In Oct. 2012 the commander of Russia's land forces, General Vladimir Chirkin, when commenting on the reintegration of the 200 Pechenga motor-rifle division into the Northern Fleet as the first 'Arctic brigade', cautioned that Russia was not rushing to establish, as announced earlier, two additional 'Arctic brigades', in order to avoid triggering a militarization of the region. See 'Главком СВ: мы с осторожностью подходим к формированию арктических бригад' [Chief Commander of Ground Forces: We handle the formation of Arctic brigades with caution], Arctic-Info, 1 Oct. 2012, <http://www.arctic-info.ru/News/Page/glavkom-sv--mi-s-ostorojnost_u-podhodim-k-formirovaniu-ar kticeskih-brigad>.

[65] Russian Ministry of Foreign Affairs (note 4).

[66] Zagorsky, A., 'Security in the Arctic: Cooperation, not competition', *Arctic Herald*, no. 1 (2014), pp. 80–85.

states.[67] The establishment of an Arctic Coast Guard Forum as well as the institutionalization of multilateral Arctic SAR exercises would be welcome initiatives from that perspective, if not affected by the Ukraine crisis.

The crisis in Ukraine and the deterioration of general relations between Russia and the West have particularly affected the search for an appropriate platform for security cooperation. Further, attempts to transcend the currently fragmented regional security architecture have been suspended or put on hold as a result. However, the impact of the Ukraine crisis on security in the Arctic is unlikely to cause an arms race but it is likely to result in lost opportunities for expanding cooperation, thus cementing the existing fragmentation.

## VII. The Arctic Council

As the Arctic region has emerged as an important issue on the international agenda, regional cooperation has become more prominent within Russian policy. This is supposed to reflect the special responsibility of Arctic—and particularly coastal—states, which exercise sovereignty and sovereign rights over large parts of the marine Arctic. On these grounds, 'decisions on Arctic matters' are supposed to be taken by Arctic countries, meaning by members of the AC. Other non-Arctic countries are supposed to follow their leadership and decisions.[68] This logic implies that, departing from their sovereign rights, it is Arctic states that have the right 'to determine on their own, which Arctic issues shall be dealt with by themselves, at the national level, which at the regional level, and which require broader international cooperation'.[69]

In terms of governance, regional frameworks have two important functions. First, they are supposed to reconfirm sovereignty, sovereign rights and jurisdictions of Arctic states, as defined by the law of the sea. Second, regional frameworks are meant to

---

[67] [Strategy for developing the Arctic Zone of the Russian Federation and providing national security until 2020. Approved by the President of the Russian Federation on 20 Feb. 2013] (note 22), para. 17.

[68] Russian Ministry of Foreign Affairs (note 4).

[69] Vassiliev (note 35).

institutionalize regional exclusiveness and shield regional decision making, to the extent possible, from non-regional influences.

Over the past two decades or so, Russia has sought to promote different regional cooperative platforms. Those included, in particular, the Barents Euro-Arctic Council (BEAC), established in 1993; the AC, established in 1996; and meetings of the coastal states (the 'Arctic Five'), which took place twice at the ministerial level (in Ilulissat, Greenland, in 2008 and in the Canadian Chelsea in 2010).

The latter format was, and still is, perceived as being important for addressing issues of specific interest for coastal states, such as the extension of the continental shelf. The Arctic Five was also conceptualized in Russia as an eventual driver of decisions to be taken by the AC on matters of particular importance for coastal states, and was expected to develop in parallel with the AC.[70] The Arctic Five continues to operate, although no longer at the ministerial level, as exemplified by the consultations of five states concerning international fisheries in the central part of the Arctic Ocean, which began in 2010 and resulted in the adoption of the Oslo Declaration in 2015.[71]

After the ministerial meetings of the Arctic Five were criticized for their exclusiveness and were discontinued after 2010, and due to the relatively narrow geographic extent of BEAC's area of activities, the AC is increasingly seen in Russia as the 'key and effective intergovernmental institution' addressing, in particular, regional issues of environmental protection and sustainable development.[72] Asserting the centrality of the AC among other regional forums, Russia proceeds on the basis that the main institutional architecture for effective Arctic governance is already in place, meaning

[70] Стенограмма ответов Министра иностранных дел России С.В.Лаврова на вопросы российских СМИ по итогам министерской встречи прибрежных арктических государств, Челси, Канада, 29 марта 2010 года [Transcription of responses by the Minister of Foreign Affairs of Russia S. V. Lavrov to the questions by Russian mass media on the outcome of the ministerial meeting of the Arctic coastal states, Chelsea, Canada, 29 Mar. 2010], <http://www.mid.ru/bdomp/ns-dos.nsf/45682f63b9f5b253432569e7004278c8/432569d8 00223f34c32576f60027ca3c!OpenDocument> (in Russian).

[71] Ryder, S., 'The Declaration Concerning the Prevention of Unregulated High Seas Fishing in the Central Arctic Ocean', 31 July 2015, <http://ablawg.ca/2015/07/31/the-declaration-concerning-the-prevention-of-unregulated-high-seas-fishing-in-the-central-arctic-ocean/>; and Zagorsky (note 47), p. 21.

[72] Vassiliev (note 35).

this architecture does not need to be altered substantially but, instead, should be consolidated and further strengthened.[73]

This conclusion leads Russia to pursue the policy of strengthening the AC along different avenues. The longer-term vision is that of gradually transforming the AC from being a high-level intergovernmental forum into a fully fledged international organization, meaning one based on a legally binding treaty.[74] Establishing a permanent secretariat for the AC in 2013 is seen as a first step in that direction.

Russia is aware that the prospects for transforming the AC into a treaty-based international organization are remote, and therefore appreciates its design as a consensus-based high-level intergovernmental forum. Instead, Russia is seeking to expand the AC's role, in particular, by advancing legally binding agreements to be developed and adopted by member states. The 2011 SAR Agreement and the 2013 Oil Spill Agreement are seen as important milestones on the AC's road to becoming a decision-making institution.

While the mandate of the AC to deal primarily with issues of sustainable development and environmental protection is currently not debated, the flexible nature of the AC as a framework facilitating broader regional cooperation beyond its direct mandate is appreciated by Russia. Meetings of chiefs of defence in 2012 and 2013, for instance, were formally held outside of the framework of the AC, but the eight member countries were present and began by discussing main avenues for cooperation in implementing the SAR Agreement.

At the same time, the AC framework is not seen as a universal platform for dealing with the whole variety of issues on the Arctic agenda. Its area of competence is largely limited by the extent of jurisdictions of its member states. It ends where the legitimate rights of other states begin. For that reason, different issues are addressed by Russia in different forums. For instance, the Polar Code establishing mandatory rules of navigation in polar ice-covered waters was developed within the IMO and not the AC.

---

[73] Vassiliev (note 35).

[74] 'Арктический совет становится межгосударственной организацией' [The Arctic Council becomes an interstate organization], Arctic-info, 15 May 2013, <http://www.arctic-info.ru/News/Page/arkticeskii-sovet-stanovitsa-mejgosydarstvennoi-organizaciei>.

Also, international fisheries in the central Arctic Ocean or the extension of the continental shelf of coastal states were neither discussed within the AC nor otherwise in a framework of its eight member states.

While promoting cooperation among its member states, the AC is not supposed to serve the purpose of transferring their sovereign rights to a regional institution, should one be established. Neither is the AC supposed to open the door to decision making for external actors. On the contrary, it is meant to keep the latter out to the extent possible, unless the relevant issues are subject to negotiations in wider institutions. The previous practices of admitting (or not admitting) new observers and defining their status within the AC are a good example of the role attributed to the AC in shielding the region from external penetration.

## VIII. Conclusions

### Back to the concept of subsidiarity

Respect for national sovereignty, sovereign rights and jurisdictions in the marine Arctic is seen by Russia as an indispensable element of any governance architecture in the region. Those rights are derived from the relevant provisions of the law of the sea and particularly from those of UNCLOS or customary maritime law. Those legal norms substantiate the unique position and privilege of coastal states and members of the AC in the governance of the region.

The sovereign rights of coastal states presume the primacy of national jurisdiction over the utilization of living resources and other economic activities within the EEZs, as well as over the exploration and exploitation of mineral resources on the continental shelf, thus determining ownership of those resources in most parts of the Arctic. Protection of sovereignty and sovereign rights is the most important mission of Russian defence forces.

Although regulation of navigation in the EEZs is not exactly a sovereign right of coastal states but one of the freedoms of high seas, coastal states are allowed by UNCLOS to extend their *environmental jurisdiction* to ice-covered waters within their EEZs.

Russia remains hesitant to abandon this right for the sake of establishing a global navigation regime throughout the Arctic Ocean.

Asserting and strengthening respect for the sovereign rights of coastal states remains the main rationale of expanding *regional governance* in the marine Arctic, either through the AC or ad hoc arrangements by the Arctic Five. Regional cooperation is also considered important in order to consolidate the ownership of large parts of the marine Arctic by regional states, and to oppose eventual claims from outside the region. Politically and legally it is seen as a means of balancing or reducing the damage from an eventual globalization or internationalization of Arctic affairs.

At the same time, evolving regional institutions are considered important for addressing common problems which transcend the maritime boundaries of Arctic states, promoting cooperation and confidence-building particularly within the fragmented security architecture. As a result, regional cooperation gradually obtains its own dynamic and helps develop a sense of common responsibility for sustainable development in the region and the conservation of its fragile environment.

Respect for the sovereign rights of coastal states is also a prerequisite for developing *wider international instruments* governing Arctic activities in order to delineate legitimate rights and responsibilities of external actors. Among the sectorial governance issues reviewed above, those that particularly exemplify this complex relationship are developments in the Polar Code, the drafting of a fisheries agreement for the central Arctic Ocean, and the internal Russian debate over the rationale of establishing an area of common heritage of humanity in the Arctic Ocean.

Russia acknowledges the legitimate navigation rights of non-Arctic states but is still hesitant to allow the exercise of these rights to be governed exclusively by a global arrangement under consideration within the IMO. It seeks to maintain national regulation within its EEZ, which essentially has a different legal nature compared with the rules established by the Polar Code.

Russia recognizes the importance of establishing precautionary measures in order to ensure conservation of living resources in the central Arctic Ocean and realizes that this goal can be difficult to achieve without engaging interested non-Arctic states. However, it emphasizes the special responsibility of coastal states and their privilege to determine the rules.

Notably, issues that go beyond the national jurisdictions of Arctic states, such as rules of navigation or international fisheries, are usually not subject to decisions (agreements) developed within *regional institutions,* thus reflecting that the legitimate rights of third countries limit the competence of those institutions. However, all the issues are subject to consideration both within the AC and eventually the Arctic Five in order to help develop a common vision and promote it in wider international frameworks. This important role of regional institutions was explicitly pointed out in 2013, when the Kiruna ministerial meeting of the AC tasked Senior Arctic Officials with identifying 'opportunities for Arctic States to use the Council's work to influence and shape action in other regional and international fora'.[75]

The formula of subsidiarity implicit in the Russian policies concerning Arctic governance follows a simple logic: national regulation enjoys priority. Regional governance arrangements are pursued to the extent necessary to complement national rules. Wider international solutions are considered as far as they appear unavoidable. As sectorial regimes in the marine Arctic differ, practical solutions and the particular mix of national, regional and wider international instruments differ depending on the issue at hand.

Russian policies concerning particular issues of Arctic governance, however, do not simply follow the implicit logic of subsidiarity but rather result from a complex process of policy formulation. This process includes a multitude of actors, often with different or diverging interests and priorities, such as: the relevant governmental agencies; both chambers of parliament; the defence establishment; different industries (oil and gas as well as mining companies, fisheries, ship owners and others); civil society; environmental organizations; associations of indigenous peoples; and Polar research and expert communities. The involved actors certainly have a different weight in the highly centralized Russian decision-making process and often reveal an affinity to the traditionalist view of the 'Russian Arctic', but the

[75] Kiruna Declaration (note 43), p. 6.

evolving Russian Arctic policies usually reflect the balance of interests of the different groups.

# 5. North East Asia eyes the Arctic

LINDA JAKOBSON AND SEONG-HYON LEE

## I. Introduction

Interest in the Arctic has increased dramatically among non-Arctic states, especially in Asia. China, Japan and the Republic of Korea (South Korea) are the frontrunners in East Asia on all matters relating to the Arctic. They are exploring ways to effectively take advantage of the opportunities and to counter the challenges of an evolving Arctic environment. In all three countries there is a growing number of internationally respected Polar researchers, who have decades of experience of examining the atmosphere and climate change. In recent years the Arctic has also become the focus of a small number of geopolitical strategists and legal experts.

The economies of these three North East Asian countries are not only dependent on exports, but they also rely on imported resources for continued economic growth. Hence, the prospect of new sea lanes opening as a result of the melting Arctic ice has direct relevance for them (see figure 5.1). The Northern Sea Route (NSR) across the northern coast of Russia, in particular, is a focus for Asia—more so than the Northwest Passage through the Canadian archipelago (see figure 5.2). This is because the NSR is expected to become commercially viable before the Northwest Passage, as a result of the ice receding more quickly off Siberia than on average across the Arctic. The opening up of the Arctic sea passage is also of interest to North East Asian because it has the potential to provide access to new reserves of energy and other natural resources, as well as new fishing grounds.

One indication of the growing interest in the Arctic among the North East Asian countries China, Japan and South Korea was the desire to become permanent observers of the Arctic Council (AC)—a wish that was fulfilled in 2013. In each country, Arctic specialists and officials voiced a desire to be more active in the AC; behind closed doors, Chinese diplomats lobbied hard for China to become a permanent observer, while Japan and South Korea adopted more subtle approaches.

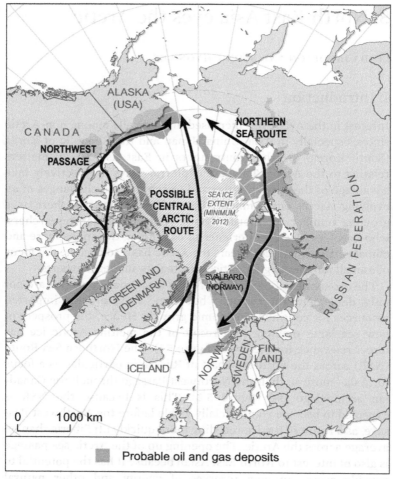

**Figure 5.1.** Arctic sea routes and potential resources

*Credit*: Hugo Ahlenius, Nordpil, <https://nordpil.se/>.

*Sources*: US Geological Survey, Circum-Arctic Resource Appraisal (CARA), 2008, <http://energy.usgs.gov/RegionalStudies/Arctic.aspx#3886223-overview>; and National Snow & Ice Data Center, 'Monthly Sea Ice Extent', 19 Nov. 2015, <https://nsidc.org/data/seaice_index/>.

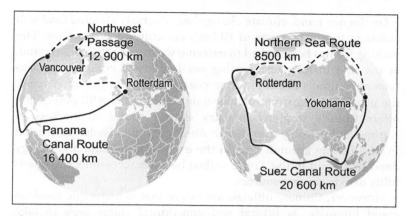

**Figure 5.2.** Comparison of the Northwest Passage and the Northern Sea Route

*Credit*: Hugo Ahlenius, Nordpil, <https://nordpil.se/>.

Sections II, III and IV provide an overview of the drivers of the Arctic interests in China, Japan and South Korea, respectively. They also review the actors that are involved in Arctic affairs and the Arctic policies that have been approved or are being contemplated in each country. Section V concludes by comparing and contrasting the interest of the three countries and raises key questions for determining the future of Arctic governance.

## II. China's Arctic activities and policies

China's interest in the Arctic region has grown significantly, especially over the past five years. Despite not being an Arctic littoral state, Chinese officials believe that the Arctic's melting ice presents both challenges and opportunities for the country's economic growth.[1]

---

[1] Jakobson, L. and Peng, J., *China's Arctic Aspirations*, SIPRI Policy Paper no. 34, Nov. 2012; Jakobson, L., 'China wants to be heard on Arctic issues', *Global Asia*, vol. 8, no. 4 (winter 2013); and Jakobson, L. and Lee, S., 'The North-East Asian states' interests in the Arctic and possible cooperation with the Kingdom of Denmark', Report prepared for the Ministry of Foreign Affairs of Denmark, SIPRI, Apr. 2013, <http://www.sipri.org/research/security/arctic/arcticpublications/NEAsia-Arctic.pdf>.

On the one hand, climate change has adversely affected (and will continue to affect) parts of China's agricultural production. The melting ice has been linked to extreme weather in the country and, in the mid to long term, rising sea levels will compel China to relocate millions of people from coastal areas. On the other hand, the prospect of ice-free summer months along the NSR potentially offers China's shipping industry shorter routes to markets in Europe and possibly even North America. China is also interested in new fishing grounds and, in the event that mineral and energy deposits buried in the Arctic seabed become accessible, the possibility of extracting resources.

Moreover, Chinese officials are aware that geostrategic tensions could intensify, as littoral and non-littoral states seek to take advantage of opportunities in the Arctic region. Two of the eight AC member states—Canada and Russia—are presumed to have very reluctantly allowed an expansion of the AC's permanent observers in 2013. For example, nine months after the AC approved the applications of five Asian nations—China, India, Japan, South Korea and Singapore—Canada's Prime Minister, Stephen Harper, said that he had had misgivings about the rush of countries and other players joining the AC as observers: 'It was just becoming literally everybody in the world wanted to be in the Arctic Council'.[2] From the viewpoint of non-Arctic states, even more worrying was Harper's statement that the Arctic should be the domain of countries with territory there and that he would oppose efforts to grant influence to outsiders.[3]

China wants to be included in discussions about the Arctic future. Throughout 2012 and early 2013 Chinese diplomats tried to convince governments of the AC member states that it is important for China to be granted permanent observer status. Since 2010 Chinese scholars and officials have emphasised the global, not only the regional, effects of the melting ice.[4] Therefore, according to the

---

[2] Chase, S., 'Only Arctic nations should shape the North, Harper tells the Globe', *Globe and Mail*, 17 Jan. 2014.
[3] Chase (note 2).
[4] The new challenges and opportunities posed by increased access to the Arctic were discussed at a workshop in Beijing on 10 May 2012, entitled 'Chinese and Nordic Cooperation on Arctic Developments'. The workshop was organized by SIPRI and the China Center for Contemporary World Studies (CCCWS). For further details see <http://www.sipri.org/media/pressreleases/media/pressreleases/2012/arcticchinapr>.

Chinese argument, non-Arctic states also have a legitimate right to be included in the new decision-making mechanisms and structures. Key unknown variables pertaining to future Arctic governance are China's desire to lobby Arctic states as the leader of non-Arctic states and the willingness of non-Arctic states to collaborate with China.

China is also strengthening its scientific activities in the region: Chinese scientists are conducting joint research projects with counterparts in several littoral states in an effort to bolster their Arctic capabilities.

### The drivers of China's Arctic interests

China's geopolitical interest in the Arctic was sparked in 2007 when Russia deployed a small submarine to the North Pole to plant a Russian flag on the seabed. Before this event, few Chinese people outside of natural sciences and environmental studies paid attention to the Arctic. Since then, a gradual awakening has taken place among Chinese Government officials and social science researchers of the need to prepare for the day when the Arctic's sea lanes will be readily accessible to vessels, at least during the summer season.

As a result of this growing awareness, over the past five years the Chinese Government has taken steps to protect what it perceives as China's key interests in the Arctic, which are: (*a*) to strengthen its capacity to prepare appropriate responses to the effects that climate change in the Arctic will have on food production and extreme weather; (*b*) to ensure access at a reasonable cost to Arctic shipping routes; and (*c*) to strengthen its ability as a non-Arctic state to access resources and fishing waters.

The Chinese Government has increased the funding of polar research and polar expeditions. A second Chinese polar research icebreaker is being built in China and will be operational in 2016. It is expected to be able to plough through ice almost two metres thick, according to the Director of the Chinese Arctic and Antarctic Administration, Qu Tanzhou.[5] The new vessel will surpass China's

---

[5] '中国新破冰船呼之欲出' [China's new icebreaker to debut], *People's Daily*, 7 Jan. 2014 (in Chinese).

only existing icebreaker, *Xuelong* (Snow Dragon), in scientific research and ice-breaking ability.[6] The design contract, valued at more than $613 million, was signed with a Finnish company in 2012. Yet two icebreakers reflect a modest polar capacity compared to Russia and the Nordic countries. Russia, for example, has five nuclear-powered icebreakers operating along the NSR. Three new nuclear-powered ones are expected to be operational along the NSR by 2020.[7]

Despite an increase in attention paid to the Arctic, it is also important to note that the Antarctic has always been the main focus of China's polar research—an emphasis that is expected to continue. Only about one-sixth of the government's polar resources are devoted to Arctic expeditions.[8] As of early 2016 China had undertaken 32 expeditions to the Antarctic, but only 6 to the Arctic.

While Chinese industry representatives have taken only a few concrete measures to prepare for the emergence of new commercial shipping routes, since 2007 the Chinese Government has provided funding to researchers to strengthen Arctic expertise. Thus Chinese officials and academics are ahead of business representatives in their knowledge about the Arctic and they are the ones who advocate that China should take advantage of the potential opportunities in the Arctic.

China's most noteworthy objectives relate to how the country can benefit from the economic opportunities borne by the warming of the Arctic and how to offset or moderate the adverse effects that a warming Arctic will have on its food security and economy. In all analyses of the Chinese Government's policies it is worthwhile to bear in mind that the foremost, publicly stated goal of the Communist Party of China (CPC) is to maintain political stability: this means keeping the CPC in power. Economic growth and development are identified as comprising the foundation of political

---

[6] 'Finland's Aker to design China's new icebreaker', ScandAsia.com, 6 Jan. 2014, <http://scandasia.com/finlands-aker-design-chinas-new-icebreaker/>.

[7] Young, O. R., Kim, J. D. and Kim, Y. H. (eds) *The Arctic in World Affairs* (Korea Maritime Institute and East-West Center: Seoul and Honolulu, Dec. 2012), p. 9.

[8] Chinese State Oceanic Administration official, Interview with authors, Beijing, Mar. 2014; and Brady, A-M., 'Polar stakes: China's polar activities as a benchmark for intentions', *China Brief*, vol. 12, no. 14 (19 July 2012). Brady stated that one-fifth of China's polar resources were devoted to the Arctic.

stability. There is a consensus among Chinese scientists that climate change in the Arctic has impacted on China's climatic conditions, its ecosystem and, subsequently, its agriculture. In 2008, for example, when China's southern city of Guangzhou recorded the coldest winter since 1984, Chinese experts effectively attributed it the warming of the Arctic.[9]

An underlying, but unstated, motive behind China's increasing Arctic activities is its desire to exert influence as a rising major power. However, reports that describe China's Arctic actions as 'assertive' should be read with caution—in reality, China's Arctic policies are still a work in progress.[10] Despite the spike in interest, the Arctic is not a priority for China's foreign policy.[11]

## China's Arctic actors

The Chinese Government handles Arctic and Antarctic matters jointly as polar affairs. China's polar activities are funded by several ministries and agencies under the State Council, which is China's highest governmental body, entrusted by the CPC with the day-to-day administration of the country. For example, the final decision in 2011 to build a new icebreaker was made by the State Council.

The State Oceanic Administration (SOA) is the key government body responsible for polar affairs in all aspects, from scientific research to strategic issues.[12] The SOA is a second-tier agency under the Ministry of Agriculture. Within the SOA, the Chinese Arctic and Antarctic Administration (CAA) directly manages polar affairs and is administratively responsible for China's polar expeditions.[13]

---

[9] Jakobson and Peng (note 1), p. 10.

[10] Perreault, F., *Can China Become a Major Arctic Player?*, RSIS Commentaries, no. 073/2012, S. Rajaratnam School of International Studies, Nanyang Technological University (RSIS: Singapore, 24 Apr. 2012); and Campbell, C., 'China and the Arctic: objectives and obstacles', US–China Economic and Security Review Commission Staff Research Report, 13 Apr. 2012.

[11] Jakobson and Peng (note 1), p. vi.

[12] Qu, T. et al. (eds), '北极问题研究' [Research on Arctic issues] (Ocean Press: Beijing, June 2011), p. 364 (in Chinese).

[13] For further information see the Chinese Arctic and Antarctic Administration website, <http://www.chinare.gov.cn/en/>.

The Chinese Advisory Committee for Polar Research (CACPR) serves as an important and active governmental coordinating body on polar issues. The CACPR is comprised of experts from 13 Chinese ministries or bureaus under the State Council and the General Political Department of the People's Liberation Army (PLA).[14] As of March 2014 it had convened at least fifteen times since its establishment in 1994.

The Ministry of Transport (MOT) oversees and regulates China's domestic and international shipping industry. The Shipping Department of the MOT directly administers China's shipping ports, routes and other facilities and is in charge of China's international shipping cooperation.[15]

The Ministry of Foreign Affairs (MFA) is officially the lead organization on issues regarding international Arctic cooperation.[16] Within the MFA, the Department of Law and Treaty prepares statements on China's official position on the Arctic, coordinates China's representation at AC ministerial meetings, and is the Chinese counterpart in bilateral and multilateral engagement between China and other states, both Arctic and non-Arctic.[17] In the MFA, an assistant foreign minister is the highest-ranking official to have elaborated on Arctic issues.

There are anecdotal indications that China's senior leaders are beginning to raise the public profile of the Arctic, although this usually takes place within the realm of general polar affairs. For example, in 2014 the Xinhua news agency mentioned a symposium with scientists researching polar regions in a report about a visit by Vice Premier Zhang Gaoli, a member of the Communist Party's top

---

[14] The 13 State Council agencies are: the Ministry of Foreign Affairs; the National Development and Reform Commission; the Ministry of Education; the Ministry of Science and Technology; the Ministry of Industry and Information Technology; the Ministry of Finance; the Ministry of Land and Resources; the National Health and Family Planning Commission; the Chinese Academy of Sciences; the China Earthquake Administration; the China Meteorological Administration; the National Natural Science Foundation of China; and the National Administration of Surveying, Mapping and Geoinformation. Qu et al., eds (note 12), p. 365 (in Chinese).

[15] For further information see the Chinese Ministry of Transport website, <http://www.mot.gov.cn/zizhan/siju/shuiyunsi/jigouzhineng/> (in Chinese).

[16] Qu et al., eds (note 12), p. 365 (in Chinese).

[17] Ministry of Foreign Affairs official, Interview with authors, Beijing, 29 Oct. 2011.

leadership, to the State Oceanic Administration in January.[18] In October, Premier Li Keqiang, in a meeting with his Finnish counterpart, was quoted as stating that 'China appreciates Finland's open attitude towards China's participation in the Arctic cooperation'.[19] In November, on a visit to Australia, China's President Xi Jinping and his wife Peng Liyuan boarded China's polar icebreaker *Xuelong*, which was docked at Hobart, Tasmania before heading to the Antarctic. China's state media showed them touring the vessels and addressing the crew, reportedly the first visit ever by China's top leader to an icebreaker.[20] The increased focus on the maritime sphere by senior officials more generally is also raising the profile of the polar regions.

## Major research institutions

The Polar Research Institute of China (PRIC), administered by the SOA, is China's principal research institution focusing solely on polar affairs. The China Institute for Marine Affairs (CIMA), under the SOA, is the core institution for Chinese research on maritime policy, legislation, and economic interests.[21] The Chinese Academy of Science (CAS) is the country's key academic and research institution for natural sciences, technological science and high-tech innovation. Within CAS, several institutes conduct scientific studies on the Arctic environment and climate change, such as the Institute of Oceanology.

---

[18] 'Chinese leader calls for more marine power', Xinhua, 26 Jan. 2014, <http://news.xinhuanet.com/english/china/2014-01/26/c_133076098.htm>.

[19] Chinese Ministry of Foreign Affairs, 'Li Keqiang meets with Prime Minister Alexander Stubb of Finland', 17 Oct. 2014, <http://www.fmprc.gov.cn/mfa_eng/zxxx_662805/t1202433.shtml>.

[20] Chinese Arctic and Antarctic Administration, '习近平登上"雪龙"号科考船 慰问中澳科考人员' [Xi Jinping boarded the research vessel 'Snow Dragon' in Australia and greeted the expedition staff], 18 Nov. 2014, <http://www.chinare.gov.cn/caa/gb_news.php?modid=01002&id=1500>; and '解局 | 习近平"极地议程"背后有什么门道' [Decoding Xi Jinping's polar agenda and its implications], 20 Nov. 2014, <http://chuansong.me/n/925734> (in Chinese).

[21] China Institute for Marine Affairs, '海洋发展战略研究所简介' [Introduction to the Ocean Development Strategy Institute], 3 June 2010, <http://www.cima.gov.cn/_d270421662.htm> (in Chinese).

*Commercial actors*

1. *Shipping companies.* Transiting the NSR north of Russia from Shanghai to Rotterdam would shorten the trip by about 2800 nautical miles (nine days' sailing time) compared to the route via the Strait of Malacca and the Suez Canal.[22] Financial savings associated with using this shorter route are estimated at about $600 000 per vessel.[23] In 2012 the icebreaker *Xuelong* was the first Chinese vessel to successfully navigate the NSR into the Barents Sea, returning to the Bering Strait via the North Pole.[24] The Director General of PRIC, Huigen Yang, noted that the trip aroused the interest of the Chinese shipping industry in the commercial viability of the Arctic route.[25]

China's enthusiasm for the potential for Arctic shipping gained new momentum in September 2013 when the China Ocean Shipping (Group) Company (COSCO), one of China's top 100 government-controlled conglomerates, successfully conducted a test run of the NSR. The voyage of COSCO's *Yong Sheng* from Dalian to Rotterdam through the NSR was nine days shorter than the conventional routes.[26] However, enthusiasm for the NSR waned two months later when COSCO's Executive Director, Xu Minjie, resigned following an investigation by the Chinese authorities for financial irregularities.[27] Given the fact that state-owned enterprises dominate the shipping sector in China, the turbulence in COSCO's senior leadership, coupled with the net losses it reported, dealt a blow to China's Arctic shipping prospects in the near term.[28]

---

[22] Maritime Safety Administration of the People's Republic of China, ' 《北极(东北航道)航行指南》7月发布' [Guidance on Arctic navigation in the Northeast Route to be published in July], State Council of the People's Republic of China, 20 Jun. 2014, <http://www.gov.cn/xinwen/2014-06/20/content_2704789.htm> (in Chinese).

[23] Ho, J., *Opening of Arctic Sea Routes: Turning Threat into Opportunity*, RSIS Commentaries, no. 101/2011, S. Rajaratnam School of International Studies, Nanyang Technological University (RSIS: Singapore, 12 July 2011).

[24] Pettersen, T., 'China starts commercial use of Northern Sea Route', Barents Observer, 14 Mar. 2013, <http://barentsobserver.com/en/arctic/2013/03/china-starts-commercial-use-northern-sea-route-14-03#.UUbz4jZHBW8>.

[25] Pettersen (note 24).

[26] Whitehead, D., 'Chinese cargo ship reaches Europe through Arctic shortcut', CCTV.com, 12 Sep. 2013, <http://english.cntv.cn/program/china24/20130912/101621.shtml>.

[27] Toh Han Shih, 'China Cosco director quits amid probe', *South China Morning Post*, 9 Nov. 2013.

[28] Zhong Nan, 'Arctic trade route opens', *China Daily*, 10 Aug. 2013.

Chinese estimates of the viability of the NSR are unreliable—as are those made in other countries. By 2020, according to one Chinese estimate, 5 to 15 per cent of China's total international trade could pass via the NSR.[29] Ten per cent of China's trade is projected to be valued at $683 billion in 2020.[30] However, expectations of the commercial viability of the route vary and projections are often inflated. For example, in September 2012 a Chinese Government official attending the 15th EU–China Summit said that 30 per cent of cargo between China and Europe is expected to transit via the NSR in the future.[31] Yet despite the potential of the NSR, it could prove commercially unprofitable for shipping companies, at least in the short term, due to high insurance premiums, lack of infrastructure and harsh operating conditions. This uncertainty is presumably the reason why China's largest state-owned shipping companies, such as COSCO, have predominantly adopted a wait-and-see approach to the Arctic.

A significant Arctic-related shipping development for China was the leasing of North Korea's Rajin Port by Hunchun Chuangli Haiyun Logistics Limited Company in China's north-eastern province of Jilin. The company is private but the lease was agreed 'in cooperation with six Chinese ministries and the Jilin provincial government'.[32] In 2008 a lease was signed for pier 1 at the port for 10 years.[33] This agreement granted China access to the Sea of Japan for the first time since 1938. Although the Arctic was not mentioned in the media reports, Chinese analysts view Rajin as a potential Arctic hub and believe that 'the opening of Arctic shipping routes will significantly add advantages to the Tumen River area'.[34] In late 2011 the lease was extended for another 20 years. A

---

[29] Vidal, J., 'Melting Arctic ice brings hope to Russian city', *Japan Times*, 2 Feb. 2014. The article refers to an estimate by the Polar Research Institute of China (PRIC).

[30] 'China plans first commercial trip through Arctic shortcut this year', *South China Morning Post*, 13 Mar. 2013.

[31] Danish Government official, Interview with authors, Beijing, Mar. 2013.

[32] Hunchun Chuangli Haiyun Logistics Limited Company representative, Telephone interview with authors, 15 Oct. 2012; and Hunchun Chuangli Haiyun Logistics Limited Company, '公司介绍' [Company introduction], <http://vip.sol.com.cn/SOL04110475> (in Chinese).

[33] Qian, H., '破解图们江困境' [Tackling the Tumenjiang dilemma], *Liaowang Dongfang Zhoukan*, no. 19 (May 2012) (in Chinese).

[34] Zhang, X. et al., '北极航线的海运经济潜力评估及其对我国经济发展的战略' [The economic estimate of Arctic sea routes and its strategic significance for the development

year later, Hunchun Chuangli's parent company, the Dalian-based Chuangli Group, leased piers 4, 5 and 6 at the port for 50 years.[35] In February 2014 the South Korean Government also dispatched a team to Rajin, including Hyundai Merchant Marine, a South Korean logistics company providing worldwide container shipping services, amid media reports that the city could be a Rotterdam in East Asia.[36]

2. *Resource companies.* Given the resource deposits in the Arctic and the managerial and technological expertise required to operate in the region's harsh conditions, China's interest in the Arctic needs to be viewed in the framework of the government's broader 'going out' strategy. Since the late 1990s the Chinese Government has encouraged both public and private sector enterprises to invest overseas, in an effort to: (*a*) acquire advanced technology; (*b*) gain managerial and international experience; (*c*) secure access to resources and commodities; and (*d*) secure a foothold in overseas markets for Chinese exports.[37] The strategic goal is to improve the international competitiveness of Chinese enterprises and—through acquisitions, joint ventures and equity holdings in foreign companies—to ensure stable and continuous access to the resources required to fuel China's economic growth. Opportunities arising from the melting Arctic ice need to be assessed in light of the growing global interests of Chinese companies even though the Arctic is very low on the agenda of any 'going out' strategy.

To date, Chinese state-owned enterprises (SOEs) have only had modest success in investing and setting up operations within the Arctic. As in so many instances involving large-scale Chinese

of the Chinese economy], *Zhongguo Ruankexue*, Zengkan no. 2 (29 Oct. 2009), p. 92 (in Chinese). The article is co-authored by five Chinese specialists, including Zhang Xia and Guo Peiqing. North Korea's Rajin port is located on the east coast of the Sea of Japan where it borders China and Russia. It is the Tumen River's final port before it meets the Sea of Japan.

[35] Ji, H., '中朝贸易加速 朝鲜开放罗津港后再开放清津港?' [Sino–DPRK trade speed up: Chongjin after Rajin?], *Diyi Caijing Ribao*, 8 May 2012 (in Chinese).

[36] Foster-Carter, A., 'North Korea's Rajin as Rotterdam? A little less crazy now', *Wall Street Journal*, 10 Feb. 2014, <http://blogs.wsj.com/korearealtime/2014/02/10/north-koreas-rajin-as-rotterdam-a-little-less-crazy-now/>.

[37] '更好地实施 "走出去" 战略' [Improving the implementation of the 'going out' strategy], Central People's Government of the People's Republic of China, 15 Mar. 2006, <http://www.gov.cn/node_11140/2006-03/15/content_227686.htm> (in Chinese).

investment abroad, China's perceived intentions in the Arctic and the approaches relied on by Chinese companies have caused controversy.

**China's Arctic policies**

China has not published an 'Arctic Strategy' nor is it expected to do so in the coming decade: the Arctic is simply not sufficiently high on its political agenda. As a non-Arctic state, China must rely on diplomatic cooperation and the positive impact of scientific engagement and investments to promote its interests in the Arctic. In the short term, ensuring access for Chinese vessels to the Arctic shipping routes at a reasonable cost will be a priority simply because the melting ice will permit regular ship transits sooner than resource exploration and extraction. This means that China will be dogmatic in emphasizing the rights of non-Arctic states when issues such as search-and-rescue requirements, environmental standards and icebreaker service fees are decided.

Additionally, the economic competitiveness of the NSR depends on the development of needed infrastructure, the progressive alleviation of technical constraints limiting navigation and the setting of appropriate Russian tariff policies. Changes in the legal framework and fee structure along with climate change may make the NSR more competitive.

*China and the Arctic Council*

Chinese Arctic specialists within both government and academia have expressed concern that the AC member states are the sole decision-makers for the region.[38] They view this as an inadequate governance structure given the global consequences of the melting

---

[38] Liu, Z., Chinese Assistant Foreign Minister, 'China's view on Arctic cooperation', Speech on the High North Study Tour 2010, Chinese Ministry of Foreign Affairs; Chen, Y., Tao, P. and Qin, Q., '北极理事会与北极国际合作研究' [The Arctic Council and Arctic international cooperation], *Guoji Guancha*, vol. 112, no. 4 (2011), pp. 17–23 (in Chinese); Wang, C., '论北极地区区域性国际制度的非传统 安全特性-以北极理事会为例' [On the characteristics of non-traditional security of regional international institutions in the Arctic region: a case study of the Arctic Council], *Zhongguo Haiyang Daxue Xuebao*, no. 3 (2011), pp. 1–6 (in Chinese); and Sun, K. and Guo, P., '北极理事会的改革与变迁研究' [Research on reform and transformation of the Arctic Council], *Zhongguo Haiyang Daxue Xuebao* (Shehuikexue Ban), no. 1 (2012), pp. 5–8 (in Chinese).

Arctic ice.³⁹ As a permanent observer in the AC, China automatically has the right to attend AC meetings, but it does not have voting rights. To secure what it perceives as its rights, China wants to see a 'globalization' of the polar region. In the words of the Director of the Chinese Arctic and Antarctic Administration, Qu Tanzhou, 'Arctic resources ... will be allocated according to the needs of the world, not only owned by certain countries ... We cannot simply say that this is yours and this is mine'.⁴⁰

## China's relations with Arctic littoral states

Caution is necessary in any assessment of China's relations with Arctic states. In the bilateral relationships that China has with the eight AC member states it would be an overstatement to claim that the Arctic is the dominant factor—possibly with the exception of Iceland. China–Russia ties and China–United States ties are extremely complex, intertwined with deep strategic, political and economic objectives. Although devoid of strategic importance, China also views its relations with the Nordic countries and Canada through multiple lenses, including lessons to be learned and advantages to be gained in diverse sectors ranging from energy and the environment to civil society and social welfare. As the Arctic is a peripheral issue, major political decisions, events and trends unrelated to it predominantly impact China's relations with the AC member states. China's foreign policy is continuously moulded by fluctuations in regional affairs and major power politics—and sometimes by outright standoffs, as in the case of repercussions from the Ukraine crisis.

China is wary of Russia's intentions in the Arctic. This reflects an underlying mix of mutual apprehension and suspicion about the other's intentions generally within China–Russia bilateral ties—despite the rhetoric by senior leaders that the countries are

---

³⁹ Liu (note 38); Cheng, B., '北极治理机制的构建与完善: 法律与政策层面的思考' [Construction and improvement of Arctic governance: thinking from legal and policy perspectives], *Guoji Guancha*, no. 4 (2011), pp. 1–8 (in Chinese); Qu et al., eds (note 12), p. 272; and Sun and Guo (note 38).

⁴⁰ Vanderklippe, N., 'For China, north is a new way to go west', *Globe and Mail*, 19 Jan. 2014.

presently enjoying their best relations in history.[41] In private conversations, Chinese officials have expressed concern that Russia will impose unreasonable fees for the use of obligatory icebreaker and search-and-rescue services in its territorial waters and exclusive economic zones (EEZs) along the NSR. This concern is shared by other Asian countries that, like China, are hoping to utilize an ice-free summer shipping season in the Arctic.

In principle, China and Russia are ideal partners in the energy sphere, considering their geographic proximity and near perfect supply and demand complementarity. However, energy cooperation has progressed in 'twists and turns', in part due to the aforementioned lack of trust between the two nations.[42] China will definitely have to partner with an Arctic littoral state in order to gain access to energy and other resources in the Arctic, once exploration is possible, and Russia will most certainly need foreign investment in order to extract energy and other resources in its Arctic EEZs. So, in principle, the two countries are complementary in the Arctic too. Yet it remains to be seen whether or not there will be a meeting of minds between China and Russia, which would transform the potential for substantial Arctic energy cooperation into actual cooperation.

The China–Russia energy cooperation is, of course, an immense strategic question that also hinges on Russia's relationship with Europe—notably in the energy sphere. A China–Russia breakthrough appeared (at least initially) to have finally happened in May 2014: the privately-owned Russian gas producer Novatek signed a deal to supply the China National Petroleum Corporation (CNPC) with three million tons of liquefied natural gas (LNG) annually for 20 years, from their joint Yamal LNG project in Russia's Arctic region.[43] The agreement was signed during Russian President Vladimir Putin's visit to Shanghai. Novatek is Russia's second-largest natural gas producer and, in the deal, Novatek owns 60 per cent of the Yamal LNG project, while the CNPC owns 20 per cent. The project is one of the largest industrial undertakings in

---

[41] Jakobson, L. et al., *China's Energy and Security Relations with Russia: Hopes, Frustrations and Uncertainties*, SIPRI Policy Paper no. 29 (SIPRI: Stockholm, Oct. 2011), p. vii.

[42] Jakobson et al. (note 41), p. 26.

[43] 'Novatek and China's CNPC sign LNG supply deal', *Moscow Times*, 20 May 2014.

the Arctic and it aims to utilize the emerging potential of a new Arctic maritime route to transport LNG to Asia—and to Europe. The investments have given China a firm foothold in the Russian Arctic.[44]

As for cooperation with other Arctic states, China has accepted invitations to participate in bilateral Arctic dialogues and exchanges with Canada and each of the Nordic countries.[45] These have mostly focused on scientific collaboration. In 2010 China and the USA began holding an annual dialogue on the law of the sea and polar issues as a part of the US–China Strategic and Economic Dialogue. However, the Arctic remains a marginal issue in these discussions.[46]

China has excellent relations with nearly all of the Nordic countries. Even in the case of Norway, which China criticized severely after the Nobel Peace Prize was awarded to jailed dissident Liu Xiaobo in 2010, relations were finally showing signs of improvement in 2015. As the Arctic ice continues to melt, ties with Nordic countries can be expected to receive additional attention and resources. The Nordic countries are all eager to strengthen Arctic cooperation with China. In 2012 the Danish Ministry of Foreign Affairs commissioned an independent study focusing on the ways in which the Kingdom of Denmark could deepen cooperation with the North East Asian countries on Arctic issues.[47] Iceland, an AC member state, became the first European country to sign a free trade agreement (FTA) with China.[48] On the one hand, the FTA is part of China's all-around effort to 'open the gate to the Arctic'; on the other hand, the FTA is part of Iceland's desire to increase its exports (e.g. seafood) to China and to attract Chinese investment. In 2014 the Icelandic Minister for Foreign Affairs and External Trade of Iceland, Gunnar Bragi Sveinsson, said Iceland was seeking

---

[44] Total, Yamal LNG', <http://www.total.com/en/energies-expertise/oil-gas/exploration-production/projects-achievements/lng/yamal-lng?%FFbw=kludge1%FF>. The French company Total owns 20% of the project.

[45] Ye, J., 'China's role in Arctic affairs in the context of global governance', *Strategic Affairs*, vol. 38, no. 6 (Nov. 2014), p. 916; and Jakobson and Peng (note 1), p. 20.

[46] Jakobson, L., *Northeast Asia Turns Its Attention to the Arctic*, NBR Analysis Brief (National Bureau of Asian Research: Seattle, WA, 17 Dec. 2012).

[47] Jakobson and Lee (note 1).

[48] The Icelandic Parliament announced the passing of the agreement in Jan. 2014. 'Icelandic parliament passes China–Iceland FTA', *China Daily*, 30 Jan. 2014.

closer partnership with China and the China National Offshore Oil Corporation (CNOOC), the country's largest offshore oil and gas developer, became the first Chinese firm licensed to look for oil in the Arctic.[49]

## III. Japan's Arctic activities and policies

The NSR presents economic opportunities and strategic challenges for Japan's leaders. Japan is well positioned to take advantage of new shipping routes because of its several large northern ports. Deposits of natural resources in the Arctic are also of interest to Japanese policymakers and commercial actors alike, particularly since the Fukushima nuclear disaster of 2011. With the use of nuclear power being scrutinized more closely than ever before, the demand for oil and gas as alternatives to nuclear power plants has increased in Japan.[50] Thus the changing Arctic environment offers the potential to invigorate the Japanese economy.[51] At the same time, the melting Arctic ice also presents Japan with new security challenges because new sea lanes could leave Japan vulnerable to a military offensive from the north. Japan has territorial disputes with Russia over the Kurile Islands, which are situated along the NSR, and this is an ongoing obstacle in bilateral relations. The disagreement has prevented Japan and Russia from concluding a formal World War II peace treaty.

### The drivers of Japan's Arctic interests

In August 2012 a major Japanese newspaper, *Yomiuri Shimbun*, described Japan as a 'latecomer' to Arctic affairs, voicing concerns that it could be left behind by neighbouring countries, particularly China, in taking advantage of the Arctic's commercial and strategic

---

[49] 'Interview: Iceland to work closer with China on Arctic development: Icelandic FM', Xinhua, 5 Mar. 2014, <http://news.xinhuanet.com/english/indepth/2014-03/05/c_133162639.htm>; and Du, J., 'CNOOC licensed to seek Arctic oil', *China Daily*, 4 Mar. 2014.

[50] Spross, J., 'Japan looking to replace lost nuclear power with fossil fuels', Think Progress, 14 Apr. 2014, <http://thinkprogress.org/climate/2014/04/14/3426534/japan-replace-nuclear-fossil-fuels/>.

[51] Ohnishi, F., *The Process of Formulating Japan's Arctic Policy: From Involvement to Engagement*, East Asia–Arctic Relations Paper no. 1, Center for International Governance Innovation (CIGI: Waterloo, Nov. 2013), p. 5.

opportunities.[52] Japan was the last of the three North East Asian neighbours to apply for permanent AC observer status. It was not until September 2010 that Japan formally recognised the strategic importance of the Arctic, when the Japanese Government officially launched its Arctic task force.[53]

This 'latecomer' label is, however, only partially accurate. In some respects Japan's interest in the Arctic goes back further than that of China or South Korea. In the 1990s Japanese shipping companies worked closely with Norwegian and Russian Arctic research institutes to explore the commercial potential of the Arctic, well before it began to gain global attention, according to one Japanese expert.[54] Yet when other countries started to engage more seriously in Arctic research, Japanese companies still remained sceptical of the economic potential of Arctic shipping and resources. One of the reasons for this scepticism was that early government-led studies concluded that the speed of melting of the Arctic ice, a crucial component of the discussions surrounding the opening of new sea lanes, was exaggerated and the Japanese Government subsequently shelved ongoing initiatives in the region. Japan's economic decline in the 1990s gave further cause for caution among its shipping companies.

Today, Japan is very much aware that the Arctic being the focus of intense interest globally and it does not want to be a latecomer a second time. As Japan's ambassador in charge of Arctic affairs said in early 2014, 'We want to actively participate'.[55] The Japanese Government has increased funds for Arctic research and international research collaboration. For example, Japan hosted the Arctic Science Summit Week in April 2015, which is an annual major event bringing together international organizations engaged in Arctic research.

[52] 'Japan needs to gain voice in Arctic Ocean development', *Yomuiri Shimbun*, 27 Aug. 2012.
[53] Japanese Ministry of Foreign Affairs, 'Launching of the "Arctic Task Force (ATF)"', 2 Sep. 2010, <http://www.mofa.go.jp/announce/announce/2010/9/0902_01.html>.
[54] Kotani, T., Interview and email exchange with authors, Jan. 2013.
[55] Vidal, J., 'Russian Arctic city hopes to cash in as melting ice opens new sea route to China', *The Guardian*, 1 Feb. 2014.

## Commercial actors

The lack of interest by Japan's commercial sector has limited the development of the government's Arctic policy. Only recently has the Japanese business community begun to seriously consider the potential opportunities of an ice-free Arctic and an NSR for maritime transportation: the southern route between the ports of Yokohama and Hamburg is approximately 21 000 kilometres, via the Strait of Malacca and the Suez Canal, whereas the polar route is about 8000 km shorter.[56]

According to Dr Tetsuo Kotani, an expert on Arctic affairs at the Japan Institute of International Affairs, 'As long as there is not a clear interest from the industry sector, it's very difficult for the government to promote [Arctic affairs] at the national level'.[57] Yet Japan's scarcity of natural resources has led industry to reconsider the potential of the Arctic as a prospective source of, and transport route for, LNG. Japanese industry has resumed collaboration with Canadian, Norwegian and Russian counterparts to investigate Arctic opportunities. According to Dr Shigeki Toriumi of Chuo University, 'Japan needs to cultivate a diversity of resource exporting partners and is watching the situation very carefully from the perspective of risk management'.[58] Yet the shipping industry largely remains sceptical about the commercial viability of the NSR in the immediate future.[59] As Kotani pointed out, 'Shipping companies don't make plans based on optimistic estimates. They need reality'.[60] It is apparent that the Japanese Government wants to encourage shipping companies to be more enthusiastic about the potential of the Arctic. The Ministry of Transport has established a committee to assess the benefits and risks of the

---

[56] Saito, Y., 'Japan looking to tap Arctic sea lanes', *Nikkei Asian Review*, 3 Feb. 2014.

[57] Kotani, T., Japan Institute of International Affairs, Interview with authors, Tokyo, Oct. 2012.

[58] Toriumi, S., 'The potential of the Northern Sea Route', ChuoOnline, 28 Feb. 2011, <http://www.yomiuri.co.jp/adv/chuo/dy/opinion/20110228.htm>.

[59] Verny, J., 'Container shipping on the Northern Sea Route', Presentation, International Transport Forum, Leipzig, 26–29 May 2009, <http://www.internationaltransportforum.org/2009/pdf/PrizeVerny.pdf>.

[60] Kotani (note 54).

shorter route to Europe and will seek input from shipping companies and cargo owners.[61]

## Strategic drivers

Japan is concerned that increased commercial activities and a rush for Arctic resources will be accompanied by an increased military presence, including naval operations, around its northern waters. In 2011 Hashimoto Yasuaki, of the National Institute of Defence Studies in Japan, wrote that if either Russia or the USA 'were to operate military surface vessels and submarines in the Arctic, this would constitute the deployment of military force very close to the other country's mainland'.[62] In December 2013 President Putin's announcement of Russia's intention to strengthen its Arctic command made Hashimoto's unease a reality.[63] Hashimoto wrote that while, at present, 'neither Russia nor the United States has had to worry too much about military encroachments by the other party in the Arctic, the opening up of the Arctic Ocean would threaten to destabilize the security situation in the "backyard" of the two nations'. Several Arctic states, in particular Russia, are expanding their military polar operations and capabilities. China and South Korea are devoting significant resources to their respective polar research programmes and capabilities. These developments have propelled Japan to also consider the strategic implications of the changing Arctic environment.

---

[61] Saito (note 56). When it comes to the Arctic, Japan wants to stay active. Quoted in Vidal (note 29), the Japanese ambassador in charge of Arctic affairs, Toshio Kunikata, said: 'We want to actively participate. We are researching the Arctic sea route.'

[62] Hashimoto, Y., 'Maintaining the order in the Arctic Ocean: cooperation and confrontation among coastal nations', ed. S. Yoshiaki, East Asia Strategic Review 2011 (National Institute for Defense Studies: Tokyo, May 2011), p. 66.

[63] AP, 'Putin vows to beef up Arctic military presence', Japan Times, 10 Dec. 2013. In addition, since Dec. 2014 Russia has launched a unified network of military facilities in its Arctic territories to host troops, advanced warships and aircraft as part of a plan to boost protection of the country's interests and borders in the region. For further information see 'Putin: Russia's Arctic command to become operational in December', Sputnik International, 24 Nov. 2014, <http://sputniknews.com/military/20141124/1015103274.html>.

## Japan's Arctic actors

The Ministry of Foreign Affairs of Japan was in charge of coordinating efforts to gain permanent observer status in the AC and to create the Arctic Task Force. The latter was established in 2010 in order to take a 'cross-sectoral approach' to Arctic foreign policy and related issues of international law.[64]

The Ministry of Education, Culture, Sports, Science and Technology in Japan (MEXT) is responsible for scientific research in the Arctic. In 2011 MEXT funded a six-year comprehensive Arctic climate change research programme. In 2010 Japan established the Ocean Policy Headquarters, under the Cabinet Office, with the aim of coordinating with different agencies on Arctic policy. However, various interviewees pointed out that these coordination efforts have been compromised by Japan's bureaucratic tradition, which has hampered inter-agency communication.[65]

### Other institutions

The National Institute for Polar Research has been conducting research on the upper and super upper atmosphere of the Arctic since the 1970s.[66] Japan Oil, Gas and Metals National Corporation (JOGMEG) is a quasi-government agency established through the integration of the Japan National Oil Corporation and the Metal Mining Agency of Japan. JOGMEG conducts research and invests in natural resource deposits and development overseas, including in the Arctic, with funding from leading Japanese companies. The Ocean Policy Research Foundation (OPRF), established as the Japan Foundation for Shipbuilding Advancement in 1975, is Japan's leading research institute on Arctic affairs. It began organizing conferences on the Arctic in Japan in 2010.[67] The Japan Institute of International Affairs, a prestigious think tank under the Ministry of Foreign Affairs, has also begun to pay attention to Arctic affairs in a

---

[64] Japanese Ministry of Foreign Affairs (note 53).
[65] Japanese think tank researchers and government scholars, Interviews with authors, Tokyo, 2013 and 2015.
[66] For further information see the National Institute of Polar Research website, <http://www.nipr.ac.jp/english>.
[67] Akimoto, K., Senior Research Fellow, Policy Research Department, Interview with authors, 26 Jan. 2011.

sign that the Arctic is rising on the agenda of Japan's policy-makers.[68]

## Japan's Arctic policies

Scepticism from the private sector presents a challenge to the Arctic ambitions of the Japanese Government. Hesitant private shipping companies must be convinced that the region is worth their time and investment, with industry experts estimating that it may be 10 years before commercial natural gas shipping via the NSR begins.[69]

### Scientific research

In August 2008 the Japanese Government published an interim report on the country's current and future Arctic strategy, focusing mainly on the scientific aspects of climate change in the region. An emphasis on science and technology has been a common strategy among non-Arctic states seeking permanent observer status with the AC. As has been the case in China and South Korea, official documents from the Japanese Government have tended to highlight scientific research objectives and to downplay the interest in natural resources, shipping and governance regimes. However, while Chinese Government officials have publicly claimed that China is an 'Arctic stakeholder', the Japanese Government has employed a more modest rhetoric of 'participation'.

### The Northern Sea Route

Japan's proximity to the Bering Strait—the entrance to the NSR—will give it an advantage over other Asian shipping hubs, including Singapore, Hong Kong, Busan in South Korea and the nascent port of Rajin in North Korea. However, Japanese experts see numerous problems with the NSR, such as the challenges of transporting time-sensitive cargo in an unpredictable ice-free period and with slower navigation through icy waters. They also suggest that the NSR could be hit by an environmental disaster, as well as stating

---

[68] Vukmanovic, O. and Koranyi, B., 'Russia Arctic natural gas shipping route to Asia 10 years away', *Insurance Journal*, 28 Jan. 2013.
[69] Vukmanovic and Koranyi (note 68).

that ships in the Arctic Ocean must run on diesel or gas, which equates to 1.5 to 2 times the units of fuel oil that ships would otherwise use on the standard southern route.[70] Moreover, ships travelling the icy NSR would have to undergo costly upgrades to their hulls. According to one Japanese expert, Japanese companies would need to hire Russian nuclear icebreakers, as building them themselves would be prohibitively expensive.[71] Transiting the Russian waters would also have additional costs: Japanese analysts note that Russia requires the use of an icebreaker vessel and the presence of an onboard pilot when traversing the NSR.[72]

*Strategic policies*

Japan believes that if a sea route through the Arctic to North East Asia becomes available, it is likely that an increased commercial presence would lead to an increased military presence. If this occurs, Japan would need to develop a new sea-lane defence strategy and improve its coastguard capabilities in order to address the increased vulnerability of its northern approaches.

Within this context, Japan's traditional regional rivalry with China is now expanding into the Arctic. More militant quarters of Japanese society have even called for Japan to discuss a defensive strategy with the USA in an effort to protect its interests in the Arctic region. In 2012 an editorial in *Yomiuri Shimbun* claimed that 'it is unavoidable that the Chinese and Russian navies will become more active in seas north of Japan. The government will need to discuss with the United States how to build up Japan's defenses against them'.[73]

---

[70] Ocean Policy Research Foundation, '日本北極海会議報告書' [Japan's Arctic Ocean Conference Report], Obtained by authors, Feb. 2013 (in Japanese).

[71] Kotani, T., Email and phone interviews with authors, Feb. and Mar. 2013.

[72] This is stipulated in a 2012 Russian law regarding vessels using the NSR. For further information see 'Climate change creates a new trade route—and new risks', Gard Insight, 23 Jan. 2014, <http://www.gard.no/Content/20738515/Gard%20Insight%20-%20Climate%20change%20creates%20a%20new%20trade%20route%20-%20and%20new%20risks.pdf>.

[73] 'Japan needs to gain voice in Arctic Ocean development' (note 52).

## Japan's relations with Arctic littoral states

Among Arctic littoral states, Japanese officials see potential for cooperation on Arctic affairs with Norway and Russia, as well as with Japan's traditional ally, the USA. Japan has imported LNG from Norway's Snohvit gas field and the LNG was shipped via the NSR.[74] Norway has also sent government delegations to Japan and engaged with Asian countries at international Arctic conferences.[75] Japan's shipping giant Mitsui O.S.K. Lines Ltd announced in 2014 that, as of 2018, the company plans to transport LNG from a gas plant to be built on the Yamal Peninsula in northern Russia.[76] If this plan materializes, it would among the first shipping companies in the world to use NSR on a regular basis.

From a geographical perspective, Russia would be a natural Arctic partner, as it owns half the Arctic coastline and the lion's share of the region's resources. As such, and as a result of the melting Arctic ice, Japanese scholars acknowledge that relations with Russia are even more important than before.[77] In December 2012 a Russian LNG tanker made the first ever Arctic winter voyage from Norway to Japan's Kita-Kyushu through the Arctic Ocean. However, Japanese scholars have reservations about increasing Japan's energy dependence on Russia, given ongoing territorial disputes between the two countries. Japan's Prime Minister, Shinzo Abe, also expressed concern over large-scale military exercises conducted by Russia not far from Japan's northern islands—although one Japanese security expert suggested that such exercises are largely aimed at China.[78]

---

[74] Bennett, M., 'LNG tanker from Norway to arrive in Japan today', Foreign Policy Association, 4 Dec. 2012, <http://foreignpolicyblogs.com/2012/12/04/lng-tanker-from-norway-to-arrive-in-japan-today/>.

[75] Research Council of Norway, 'Norway and Japan to work closely together on polar research', 24 Oct. 2013, <http://www.forskningsradet.no/en/Newsarticle/Norway_and_Japan_to_work_closely_together_on_polar_research/1253981419189?lang=en>.

[76] 'Mitsui O.S.K. to pioneer Arctic route for LNG', *Japan Times*, 9 July 2014.

[77] See e.g. 'Japan's relations with Russia are much more important in the days of the global warming', Conference material, Arctic Conference Japan, Ocean Policy Research Foundation, Tokyo, Mar. 2012.

[78] Reuters, 'Japan "strongly protests" Russia maneuvers on disputed isles', 13 Aug. 2014, <http://www.reuters.com/article/2014/08/13/us-japan-russia-idUSKBN0GD08120 140813>; and Kotani, T., Interview with authors, Mar. 2013.

While Japan has held talks with Canada on Arctic energy development, Canada's sovereignty claims over islands north of Japan's mainland remain a sticking point for Arctic cooperation. Canada claims the energy resource-rich Arctic area as part of its territories and sees the region closely linked to its national interest. Canada's decision to assertively pursue its claim to the Arctic region using both political rhetoric and legal means may, according to Japan, set an example for and motivate China to embolden its disputed sovereignty claims to its nearby seas. According to Dr Kotani, Japan's concern is that 'the Canadian claim could give legitimacy to China's behaviour in the South China Sea'.[79]

## IV. South Korea's Arctic activities and policies

The South Korean Government sees its Arctic programme as a means of enhancing the country's international profile and playing a role in global governance commensurate with its economic standing. South Korea is a resource-poor country heavily reliant on energy imports with a significant proportion of its gross domestic product (GDP) made up of exports. As the world's largest shipbuilder, industry representatives and officials in South Korea see medium- and long-term potential in Arctic resources and use of the NSR, but the current consensus is that the NSR is not commercially viable in the short term.

From the government's perspective, a key goal of the country's polar research is to utilize its permanent observer status in the AC in order to shape a new legal order governing polar affairs. As a reflection of its ambitions in the region, in July 2013 South Korea became the first Asian country to produce a comprehensive Arctic strategy: the Arctic Comprehensive Initiative.[80] The aim is to establish a unified national strategy on Arctic affairs and better facilitate intergovernmental collaboration on Arctic affairs. South Korea's overall strategic mindset towards the Arctic is evident in the

---

[79] Kotani, T., Phone interview with authors, 21 Mar. 2013.
[80] South Korean Ministry of Oceans and Fisheries (MOF), '북극 종합정책 추진계획' [The Arctic Comprehensive Initiative], <http://www.mof.go.kr/article/view.do?menuKey=376&boardKey=10&articleKey=1250> (in Korean). The MOF spearheaded the initiative in coordination with other government departments.

initiative's slogan: 'to contribute to international cooperation and safeguard national interests'.

## The drivers of South Korea's Arctic interests

In 1996 South Korea joined the Organisation for Economic Co-operation and Development (OECD), having risen from poverty in the 1950s after the Korean War.[81] Today, South Korea is the 15th largest economy in the world and is home to global companies such as Samsung and Hyundai.[82] While South Korea has traditionally paid little attention to the Arctic region, the government now hopes to exploit mid- and long-term economic opportunities opening up in the region. With the Arctic Comprehensive Initiative, it wants to develop an Arctic business model that combines the use of the NSR, natural resource development, sales of maritime technology and seafood resources. It has set up a specialized task force that focuses on Arctic-related legal affairs. In addition, it is building a second state-of-the-art icebreaker for deployment exclusively to the Arctic. The new icebreaker is expected to double the ice-breaking capacity of the current one and is to be completed by 2020.[83] The government is also seeking involvement in the AC as part of a broader effort to enhance South Korea's global profile.

### Commercial drivers

South Korea's interests in the Arctic are linked to resources and shipping.[84] In 2011 it was the world's second largest importer of LNG, fourth largest importer of coal and fifth largest importer of crude oil.[85] South Korea imports 100 per cent of its oil. Academic papers and media reports emphasize the possibility of the NSR

---

[81] '가난한 코리아'에서 스마트폰 · 車수출 "퍼스트무버" 로' [From 'poor Korea' to 'First mover': export in smartphones and cars], Herald Economy, 29 Nov. 2013 (in Korean), <http://news.naver.com/main/read.nhn?mode=LSD&mid=sec&sid1=101&oid=016&aid=0000470081>.

[82] OECD, 'OECD Economic Survey of KOREA', June 2014, <https://www.oecd.org/eco/economic-survey-korea.htm>.

[83] '북극해 전용 제2쇄빙선 건조해 2020년 취역한다' [A second icebreaker will be in use in 2020 exclusively for the Arctic], Donga Ilbo, 2 Jan. 2015 (in Korean).

[84] Brady, A-M. and Kim, S., 'Cool Korea: Korea's growing Antarctic interests', Unpublished manuscript, Used with the authors' permission, p. 32.

[85] US Energy Information Administration, <http://www.eia.gov/beta/international/analysis.cfm?iso=KOR>.

reducing prices for Arctic oil and gas in East Asia.⁸⁶ In that context, the promise of improved shipping and opportunities for extracting natural resources helps to justify the country's Arctic programme to the domestic audience. However, the economic competitiveness of the NSR depends on the development of infrastructure, the progressive alleviation of technical constraints limiting navigation, and appropriate Russian tariff policies. Changes in the legal framework and fee structure, along with climate change, could indeed make the NSR more competitive.

South Korea is not just concerned with imports via the NSR. The opening of the Arctic Ocean to international shipping is also of interest to the country's domestic manufacturing export industries. In 2013 the value of South Korean exports accounted for 56.5 per cent of the country's GDP; South Korea's economic performance is closely aligned with its ability to export manufactured goods to developed markets.⁸⁷ The NSR, in this respect, represents a significant economic opportunity for South Korean industry to shorten the shipping distance for products bound for European and North American markets.

South Korea is fully aware that the NSR's commercial viability will not be achieved in the short term. Park Jin-hee, a researcher at the Korea Maritime and Ocean University, optimistically estimates that the NSR will be fully open, year-round, for all kinds of freight shipping after 2030.⁸⁸ This distant date, however, has not dampened enthusiasm in South Korea. When Park's team published the results of a study into which South Korean port city is likely to become the nation's NSR hub in 2030, the media paid keen

---

⁸⁶ See e.g. 'China's Silk Road plans could challenge Northern Sea Route', Barents Observer, 6 Jan. 2015, <http://barentsobserver.com/en/business/2015/01/blog-chinas-silk-road-plans-could-challenge-northern-sea-route-06-01>; and Bennett, M., 'Arctic LNG: The energy on East Asia's doorstep', 15 May 2014, <http://www.isn.ethz.ch/Digital-Library/Publications/Detail/?ots591=0c54e3b3-1e9c-be1e-2c24-a6a8c7060233&lng=en&id=180379>.

⁸⁷ 'S. Korea's ratio of exports to GDP hits new record last year', Yonhap News Agency, 8 Mar. 2013, <http://english.yonhapnews.co.kr/business/2013/01/08/77/0501000000AEN20130108002900320F.HTML>.

⁸⁸ '북극 항로 시대 경쟁력 1위 항만은 부산' [The Busan port has the number one competitive edge over other port cities in the era of Northern Sea Route], Busan Ilbo, 11 Mar. 2014 (in Korean).

attention.[89] The study concluded that the port city of Busan had the best conditions to serve as the nation's Arctic hub and recommended that the government increase its financial investment there—to expand its facilities and get it ready to be 'the outpost for the Arctic'.[90] Yet it is worth noting that other coastal cities, namely western Inchon and eastern Ulsan, are also competing to become the nation's hub and fully tap into the NSR's future potential.[91] These cities see the NSR as a new avenue for boosting local economy and so local lawmakers are also very active in promoting the route. This local competition is set to further enhance awareness of the Arctic's potential in South Korea.

### International standing

The South Korean Government regards its Arctic initiative as a medium for upgrading its international status and promoting its image in global governance, commensurate with its economic standing.[92] South Korea has sought to expand its international influence in order to create, as the administration of President Lee Myung-bak (2008–13) called it, a 'Global Korea'.[93] South Korea's global outreach stems from an awareness of its increasing economic stature and pressure to play a more active role in the international community. The United Nations Secretary General, Ban Ki-moon, who was born in South Korea, has publicly denounced the country's limited international contributions relative to the size of its economy.[94]

---

[89] '북극항로 개설에 따른 국내 무역항 경쟁력 분석' [Analysis of South Korean ports' competitiveness with the advent of the NSR], *Busan Ilbo*, 11 Mar. 2014 (in Korean).

[90] [Analysis of South Korean ports' competitiveness with the advent of the NSR] (note 89).

[91] These cities have made robust efforts to host international Arctic conferences. See e.g. <http://incheonport.tistory.com/1951> (in Korean).

[92] Brady and Kim (note 84), p. 18.

[93] South Korea has hosted or co-hosted international sporting events, such as the Olympics (1988) and the World Cup (2002), and international summit meetings such as the G20 Summit (2010). After these events, South Korean media often ran long pieces examining how these events helped the country to be 'recognized' by the world.

[94] '선진그룹으로 성장한 기대에 걸맞은 역할을 해야' [South Korea has become a developed country. It should play a commensurate role to meet the expectation of the international community], UN chief Ban Ki-moon's speech in Seoul on 10 Aug. 2011, *Dong-A Ilbo*, 12 Sep. 2009 (in Korean). See similar remarks by Ban Ki-moon at <http://www.asiae.co.kr/event/afef/forum_news_view.htm?pg=1&idxno=2011081010425646202> (in Korean).

One area in which South Korea has sought to raise its international profile is green growth and climate change, including through the Global Green Growth Institute, headquartered in Seoul, and the Green Climate Fund, headquartered in Incheon.[95] South Korea also hosted the World Water Forum in April 2015. Climate change has been, and continues to be, a high political priority in South Korea.

Although interviews with foreign ministry officials tend to focus on policy aspects and the significance of South Korea's increasing role in global governance by becoming a permanent observer in the AC, South Korea's fishing sector also pays due attention to the Arctic seas in terms of expanding the nation's fishery resources. For instance, an official at the government-run Korea Maritime Institute, Lee Seong-woo, said in 2013 that the Arctic seas are a 'potential growth point for Korea's deep-sea fishery'.[96] A director at the Korea Overseas Fishers Association, Song Gi-seon, noted that the Arctic seas are closer to South Korea than Antarctica and thus provide advantages in expense reduction for fishing.[97]

## South Korea's Arctic actors

### The Government

Under the current Park Geun-hye administration, Arctic affairs generally come under the jurisdiction of the Ministry of Oceans and Fisheries (MOF), which governs the implementation of maritime shipping and fisheries policy. MOF also produced the '2015 Arctic Policy Implementation Plan' in coordination with other government ministries. With its research and development funds, MOF also supports polar infrastructure and vessels, such as the icebreaker Araon and a second Antarctic base. For Arctic topo-

---

[95] Krukowska, E., 'UN's Green Climate Fund plans headquarters in South Korea', *Bloomberg*, 21 Oct. 2012.
[96] '북극해 어업전담기구 구성 원양어업 재도약' [Formation of fishery consultation group to offer the deep fishery industry a chance to rebound], Hyundai Haeyang, 10 Oct. 2014, <http://www.hdhy.co.kr/news/articleView.html?idxno=864> (in Korean).
[97] [Formation of fishery consultation group to offer the deep fishery industry a chance to rebound] (note 96). The comments were made during the 'Forum for establishing strategy for the Arctic seas', co-hosted by the government-run Korea Maritime Institute and the Ministry of Land, Infrastructure and Transport.

graphy and sea lanes, the Ministry of Land, Infrastructure and Transport (MOLIT) has been working closely with MOF. By 2018, MOLIT plans to complete the updating of an Arctic spatial map, including one on the trend of melting ice.[98]

The Ministry of Foreign Affairs (MOFA) represents South Korea in Arctic regional forums and the Ministry of Trade, Industry and Energy (MOTIE) is a key player in Arctic resource development and energy supply. The presidential office of President Lee Myung-bak (2008–2013) consulted MOTIE prior to his September 2012 visit to Ilulissat, Greenland.[99] MOTIE was also the South Korean signatory to a memorandum of understanding with Greenland on joint geological surveys, resource exploration and technological cooperation.

In 2015 the South Korean Government put aside 82 billion won ($71 million) for polar research activities.[100]

### The Korea Polar Research Institute (KOPRI)

KOPRI is South Korea's leading research body on Arctic affairs.[101] Located in Inchon, it employs some 200 researchers and staff, principally scientists (it does not specialize in strategic or security policy). KOPRI has three polar research bases: the King Sejong Station and the Jang Bogo Station in Antarctica and the Arctic Dasan Station in Svalbard, Norway.

KOPRI is also responsible for the activities of *Araon*, a 110-metre long, high-tech, multi-purpose icebreaker. Costing $1 billion, it was completed in 2009 and has since been deployed annually to the

---

[98] In South Korea, maritime functions have been divided among various government agencies and these agencies often change their names or merge with other agencies, making it hard for non-Korean researchers to track them down. In the past, for instance, MOF had a different name: the Ministry of Maritime Affairs and Fisheries (MOMAF). In 2008, MOMAF merged with the Ministry of Construction and Transportation (MOCT) to become the Ministry of Land, Transport and Maritime Affairs (MLTM). When Park Geun-hye became the president in 2013, she split MLTM into the Ministry of Land, Infrastructure and Transport (MOLIT) and MOF. Currently, MOF takes a more leading role in Arctic affairs than other ministries.

[99] 'South Korean President Lee Myung-bak in Ilulissat', Greenland Today, 10 Sep. 2012, <http://greenlandtoday.com/gb/category/south-korean-president-lee-myung-bak-in-ilulissat-384/>.

[100] South Korean MOF, 2015, <http://www.mof.go.kr/jfile/readDownloadFile.do?FileId=MOF_ARTICLE_7095&fileSeq=1> (in Korean).

[101] For further information see the KOPRI website, <http://www.kopri.re.kr/exclude/userIndex/engIndex.do>.

Arctic on scientific research assignments. These have been in collaboration with nine other countries and with scientists from the Arctic states on board.[102]

## South Korea's Arctic policies

South Korean industry experts foresee mid- and long-term economic opportunities as the ice-free summer period increases and the NSR becomes economically viable. When the NSR is navigable, the transit distance between Rotterdam and Busan (South Korea's largest seaport) will be reduced by 37 per cent (from 20 100 km to 12 700 km) and the transit time will be reduced from 30 to 20 days.[103]

As the world's largest shipbuilder, South Korea expects changes in the Arctic to increase demands for icebreakers, oil and LNG tankers, and sea-floating plants in the medium to long term. It also sees opportunities to participate in the extraction of natural resources. South Korea sees its entrance into the Arctic as 'the expansion of economic territory'.[104]

### Arctic Council membership

Since taking office in February 2013, President Park Geun-hye has elevated Arctic affairs to a national priority in South Korea. She listed AC permanent observer status among '140 national agenda tasks' for her presidency, and gaining this status in May 2013 was seen as an important step in securing influence within Arctic affairs.[105]

In early 2014 South Korea's Arctic interest intensified when some economic strategists at government think tanks argued that

---

[102] KOPRI, 'Polaris for the future', no. 7 (2010), <http://www.kopri.re.kr/www/about/kopri_publication/annual_publication_17.pdf> (in Korean).

[103] '꿈의 북극항로 亞·유럽 운항시간 단축 부산항에 혜택' [The Arctic dream route: reducing the distance will benefit Busan], *Busan Ilbo*, 14 Nov. 2011 (in Korean).

[104] '남북극 경제영토 확장 적극 추진' [Lee Joo-young to aggressively promote the expansion of economic territories in the polar regions], Financial News, 4 Mar. 2014, <http://news.naver.com/main/read.nhn?mode=LSD&mid=sec&sid1=100&oid=014&aid=0003108058> (in Korean).

[105] South Korean Government, '박근혜정부 국정과제' [The Park Geun-hye administration's 140 national tasks], Feb. 2013, <http://www.pmo.go.kr/pmo/inform/inform01_02a.jsp> (in Korean).

the NSR should be incorporated into the 'Eurasia Initiative'. This initiative is President Park Geun-hye's ambitious plan to help integrate North Korea into the international community by linking commercial railways from the Korean Peninsula all the way to northern Europe.[106]

## South Korea's relations with Arctic littoral states

South Korea hopes to identify an Arctic littoral state with which it can form a mutually beneficial partnership. In Arctic shipbuilding, Russia and South Korea are key partners. Following a Russia–South Korea summit meeting in 2013 between Vladimir Putin and Park Geun-hye, Russia's Arctic Ambassador, Anton Vasiliev, visited South Korea and held initial consultations to promote cooperation in the Arctic.[107]

Among the smaller Arctic states, South Korea has maintained a robust working relationship with Norway. According to the MOF Administrator of the Busan Regional Office, Jeon Ki-Jung, South Korea sees Norway as a potential Arctic partner that 'plays a leading role in Arctic navigation, offshore plant operation, greenhouse gas reduction in maritime transport', and the two nations co-hosted a maritime logistics and policy conference in Oslo in November 2014.[108] South Korea operates a research base at Ny-Ålesund, Svalbard.

South Korea has also cooperated on Arctic affairs with the USA and Canada. The Arctic was on the agenda in 2013 when South Korea and the USA held the first meeting of the Environmental Cooperation Commission under the US–Korea Environmental Cooperation Agreement.

---

[106] '유라시아 철도망 연계하고 북극항로 개척해야' [Korea should link the Eurasian railways and open up the Northern Sea Route], Yonhap News Agency, 2 Mar. 2014, <http://news.naver.com/main/read.nhn?mode=LSD&mid=sec&sid1=101&oid=001&aid=0006784650> (in Korean).

[107] TASS, 'Russia and South Korea hold first consultations on Arctic cooperation', 9 Feb. 2014, <http://en.itar-tass.com/russia/718257>.

[108] '해양수산부, 노르웨이와 북극해 항로 등 해운협력 강화' [Ministry of Oceans and Fisheries to strengthen cooperation with Norway on NSR], Ajunews, 17 Nov. 2014, <http://www.ajunews.com/view/20141117100558455> (in Korean). Note: in Mar. 2013 South Korea's Ministry of Maritime Affairs and Fisheries changed its name to the Ministry of Oceans and Fisheries.

In 2011 state-run Korea Gas Corporation (KOGAS) acquired a 20 per cent stake in the Canadian northern Umiak gas reserve owned by Calgary-based MGM Energy Corporation, marking South Korea's first Arctic resource investment.[109] In 2012 South Korea's icebreaker *Araon* explored the Canadian Arctic seas to check for gas hydrate reserves in the seabed. When Canada and South Korea signed an FTA in 2014, the Canadian Prime Minister, Stephen Harper, specifically underscored 'Arctic research and development' in a joint statement with President Park Geun-hye.[110]

## V. Conclusions

China, Japan and South Korea are all eager to take advantage of the potential opportunities that the melting Arctic ice could give rise to. As these nations are dependent on foreign trade, specialists in all three are considering the pros and cons of the NSR. At present, the sceptics still outnumber the optimists regarding the NSR's commercial viability—with good reason. Before the relevant infrastructure is in place, it is hard to imagine that the short three-month summer shipping season could become a lucrative option for standard commercial shippers. There are also significant safety and navigational concerns that remain an obstacle. Today, about one million tonnes of cargo are transported along the NSR every year, which pales in comparison to the 700 million tonnes transported along the more traditional route via the Suez Canal.[111] Furthermore, the widened Panama Canal will presumably be a competitive factor in assessments of the various shipping routes.

Among the three North East Asian countries, Japan appears to be the most cautious—or, according to some, realistic—about the potential and commercial viability of the NSR. Japan is also conceivably the most concerned about the possible destabilization of the security situation in its near vicinity, as a result of Russia's

[109] 'South Koreans eye Arctic LNG shipments', *Globe and Mail*, 19 Apr. 2011.
[110] UPI, 'Canada inks free-trade deal with South Korea', 11 Mar. 2014, <http://www.upi.com/Business_News/Energy-Resources/2014/03/11/Canada-inks-free-trade-deal-with-South-Korea/UPI-76661394540407/#ixzz2viOPkLOf>.
[111] Reuters, 'Arctic route has little allure for shipping industry', *South China Morning Post*, 16 Oct. 2013.

strengthened Arctic command and possible countermeasures that China or the USA might take.

China views the Arctic via multiple lenses, in terms of: detrimental effects on agriculture (food security) and political stability (reallocation of people from coastal areas); and potentially enhancing economic growth (shipping); food security (fishing) and energy security (resources). Chinese Government officials also view the Arctic as a new, natural dimension of China's expanding global interests and ambitions as a rising power. Although the Arctic has strategic significance and China has shown a long-term strategic perspective by stating that it wants to be part of discussions regarding future Arctic governance, there is no evidence of the country harbouring military ambitions in the Arctic.

South Korea seeks to be perceived as one of the major regional powers and its policymakers, therefore, see the Arctic as a means of gaining a seat at the major power table. To this end, it aims to strengthen bilateral ties with Arctic states.

China has been the most vocal of the three about what it views as the rights of non-Arctic states to have their voices heard when new security structures and governance mechanisms are discussed and decided. Due to China's size, its increasing economic political and military power, and the uncertainty that its rise evokes, any statement by a Chinese Government official is paid attention to by governments in neighbouring countries and even further afield. No one knows with certainty how China will use its power and, throughout history, a rising power has always caused anxiety. Nevertheless, Japanese and South Korean policymakers share China's view that non-Arctic states should also be included in any new structures governing the Arctic. Ironically, the Arctic is an area that would be ideal for these states to collaborate closely on.[112] Yet due to a number of politically explosive issues unrelated to the Arctic, including historical grievances and territorial disputes in the East China Sea, Japan has tense relations with both South Korea and China.

From the viewpoint of future Arctic governance, the most significant uncertainty—and the most fundamental challenge—

---

[112] Jakobson, L., 'China prepares for an ice-free Arctic', SIPRI Insights on Peace and Security no. 2010/2, Mar. 2010, p. 13.

pertains to how the five Arctic states deal with the views and concerns of 'outsiders' like China, Japan and South Korea. These three nations are not alone in their view that non-Arctic states have legitimate interests and concerns in the maritime Arctic. Countries as far afield as Brazil, India and South Africa have interests in the Arctic, related to commercial shipping, oil and gas development, and tourism.[113]

For their part, the five coastal Arctic states have moved on from a rigid stance that only they have legitimate interests in the Arctic. However, despite an emphasis on inclusiveness—for example, in a 2010 statement by Hillary Clinton—Arctic officials in the North East Asian states continue to have concerns about the stance of particularly the larger coastal Arctic nations towards outsiders.[114] In research interviews, North East Asian officials referred to two Arctic Five (A5) meetings, in 2008 and 2010, which excluded both AC member states without an Arctic coastline (Finland, Iceland and Sweden) and member organizations representing indigenous peoples. They also referred to part of the 2008 Ilulissat Declaration, according to which the five coastal states, by virtue of their sovereign rights and jurisdiction in large areas of the Arctic Ocean, are in a unique position to address the challenges of the transforming environment.[115]

The concerns of the non-Arctic states have not been fully allayed, not even by the 2011 Nuuk ministerial meeting that reaffirmed the primacy in the AC (tacitly reaffirming the importance of all eight members states and organizations) as the principle vehicle for addressing matters of common concern in the Arctic.[116] As Oran R. Young points out: 'In effect, what had been the stance of the A5 regarding relations with the non-Arctic states has re-emerged as the stance of the A8 in a move to draw a clear

---

[113] Young, O. R., Kim, J. D. and Kim, Y. H., 'Introduction and overview', eds Young, Kim and Kim (note 7), p. 5; and Young, O. R., 'Informal Arctic governance mechanisms', eds Young, Kim and Kim (note 7), p. 282.

[114] Chinese, Japanese and Korean Arctic officials, Research interviews and discussions with authors, 2011–14. On Canada's stance towards inclusiveness see e.g Plouffe, J., 'Canada's tous azimuts Arctic foreign policy', Northern Review, no. 33 (spring 2011), pp. 73–76.

[115] 'The Ilulissat Declaration', Arctic Ocean Conference, Ilulissat, Greenland, 27–29 May 2008, <http://www.oceanlaw.org/downloads/arctic/Ilulissat_Declaration.pdf>.

[116] Young, 'Informal Arctic governance mechanisms' (note 1133), p. 281.

line between the Arctic states and the non-Arctic states in addressing issues on the Arctic agenda.'[117]

As to the question of who has the right to decide how the Arctic will be governed, it is difficult to imagine how the different expectations will be resolved. Much will depend on the manner in which the North East Asian states, in particular China, develop their relations in the Arctic sphere with the five Arctic coastal states and how they interact with all the members at the AC. It remains unclear how ambitious the governments of China, Japan and South Korea are with regard to Arctic governance issues. Will they substantially increase the number of experts sent to work in the AC's working groups? Will they bring up Arctic politics and lobby for the rights of non-Arctic states in bilateral discussions with each of the AC member states? Will they suggest establishing an alternative international body to the AC if their views are not considered?

These questions will become more pertinent as the strategic importance of the Arctic increases. The controversies surrounding the Arctic could escalate in the event that smaller Arctic states (the Nordic countries) adopt conciliatory policies towards China and other non-Arctic states (seeking partners to work with on Arctic projects), while bigger Arctic states take more standoffish positions towards China and other non-Arctic states.[118] It is also possible that China's role as the economic facilitator of massive Arctic resource projects will place it in a strong position vis-à-vis the Arctic littoral states, especially once the five states come to an agreement among themselves on issues of delimitation; how China–Russia relations progress will be pivotal in this regard. In addition, the debate about the future of the Arctic is likely to further intensify if the rift deepens between those who want to open up the region to resource exploration, requiring multinational operations, and those who are more concerned about the climatic and environmental effects.

---

[117] Young, 'Informal Arctic governance mechanisms' (note 113), p. 282.
[118] Ng, T., 'China can influence Arctic Council agenda: Danish Minister Lidegaard', *South China Morning Post*, 27 Apr. 2014.

# 6. The Arctic Council in Arctic governance: the significance of the Oil Spill Agreement

SVEIN VIGELAND ROTTEM

## I. Introduction

The Arctic Council (AC) is often referred to as being the most important international forum in the Arctic and its influence continues to grow. The establishment of a permanent secretariat in Tromsø, Norway, and the signing of the first internationally binding agreement created under the auspices of the AC have raised its political visibility in the past five years. The Agreement on Aeronautical and Maritime Search and Rescue in the Arctic (SAR Agreement) was signed at the May 2011 ministerial meeting in Nuuk, Greenland, while the permanent secretariat was officially opened in conjunction with the Arctic Frontiers Conference in the winter of 2013.[1] The signing of a new binding international agreement on oil spill preparedness and response (Agreement on Cooperation on Marine Oil Pollution, Preparedness and Response in the Arctic, hereafter the Oil Spill Agreement) at the May 2013 ministerial meeting in Kiruna, Sweden, provides further evidence that the AC is now taking a more proactive role in Arctic governance.[2] It is becoming a leading actor in the region, both as a producer of knowledge and as an arena for the drafting of binding international agreements.

---

[1] Agreement on Aeronautical and Maritime Search and Rescue in the Arctic, opened for signature 12 May 2011, entered into force 19 Jan. 2013, <http://arctic-council.npolar.no/accms/export/sites/default/en/meetings/2011-nuuk-ministerial/docs/Arctic_SAR_Agreement_EN_FINAL_for_signature_21-Apr-2011.pdf>. During the joint Nordic presidency of the Arctic Council from 2006 to 2013, it had a temporary secretariat in Tromsø. The official creation of the permanent secretariat took place in May 2013.

[2] Agreement on Cooperation on Marine Oil Pollution, Preparedness and Response in the Arctic, opened for signature 15 May 2013, <http://www.arctic-council.org/eppr/agreement-on-cooperation-on-marine-oil-pollution-preparedness-and-response-in-the-arctic/>. Five states had ratified the agreement as of Oct. 2015: Canada, Finland, Iceland, Norway and Russia.

This chapter uses the Oil Spill Agreement as a basis to examine the type of impact AC initiatives could have on Arctic governance. Although the agreement was not drawn up by the AC, it was negotiated within its framework, by its eight permanent members (Canada, the Kingdom of Denmark, Finland, Iceland, Norway, Russia, Sweden and the United States). One question raised by the agreement is how it has influenced international and national policy in its field. While it is too early to assess the full impact of the Oil Spill Agreement, this chapter examines its potential importance, and what the signing of it indicates about the balance between international and national governance in the Arctic. Norway, which was one of the driving forces behind the agreement, is used as an illustrative example for this analysis.

Thus, the aims of this chapter are two-fold. First, to explain the reality of the Oil Spill Agreement and what it might mean for future agreements. Second, to assess the balance between international and national governance in the Arctic, and where the AC fits into this framework. Before examining the substance of the agreement, it is necessary to present a brief history of the AC to establish the context in which the agreement arose and to provide detail of the AC's evolution from a collaborative entity tasked with protecting the environment in the North to an arena where states can negotiate binding international agreements.

## II. A brief history of the Arctic Council[3]

In the 1980s the strategic military rivalry between the Soviet Union and the USA dominated the political agenda in the Arctic. At the same time, awareness of the environmental problems affecting the region began to grow.[4] In his famous Murmansk speech, the then Soviet leader Mikhail Gorbachev expressed his ambition to transform the Arctic into 'a zone of peace'.[5] Similarly, the stated

---

[3] This section draws on Rottem, S. V., 'The Arctic Council and the Search and Rescue Agreement: the case of Norway', *Polar Record*, vol. 50, no. 3 (2014); and Rottem, S. V., 'A note on the Arctic Council agreements', *Ocean Development and International Law*, vol. 46, no. 1 (2015).

[4] Pedersen, T., 'Debates over the role of the Arctic Council', *Ocean Development and International Law*, vol. 43, no. 2 (2012).

[5] Gorbachev, M., Speech in Murmansk at the Ceremonial Meeting on the occasion of the presentation of the Order of Lenin and the Gold Star to the city of Murmansk, 1 Oct. 1987.

objective of Canada and the USA at the time was for the Arctic to become a region of cooperation.⁶ In September 1989 the Finnish Government called on the Arctic states to work together to protect the Arctic environment—the so-called Rovaniemi process.⁷ It was decided that the various countries' authorities with a responsibility for the Arctic environment should meet regularly.⁸ The undertaking came to be known as the Arctic Environmental Protection Strategy (AEPS).

In 1995 Canada sought to expand the AEPS into an international organization. This proposal met with resistance, however, particularly from the USA. Instead, the Arctic governments agreed to organize their cooperative efforts in the shape of a forum and in 1996 the AC was officially established through the Ottawa Declaration.⁹ The AC is a forum without a legal personality and therefore is not an international organization as such.¹⁰

## The Arctic Council's decision-making structure in outline

The statutes governing the AC were decided at the ministerial meeting in Iqaluit, Canada, in 1998.¹¹ A key requirement is that all decisions in the forum and subordinate working groups are to be made by consensus.¹² The actual work of the AC proceeds on three levels: ministerial, senior civil servant (Senior Arctic Officials, SAOs) and working group. When the AC states hold ministerial meetings, which constitute the highest decision-making authority and usually occur every second year, such gatherings tend to

---

⁶ Pedersen (note 4).
⁷ Pedersen (note 4).
⁸ The 8 circumpolar countries are Canada, the Kingdom of Denmark, Finland, Iceland, Norway, Sweden, Russia and the USA. A total of 5 states have coastal rights in Arctic areas: Canada, the Kingdom of Denmark, Norway, Russia and the USA.
⁹ In addition to the 8 Arctic states, a number of indigenous organizations have status as permanent members of the Arctic Council. Several states and organizations are also accredited observers. For information on the organizational structure see the Arctic Council's website at <http://www.arctic-council.org>.
¹⁰ Bloom, E. T., 'Establishment of the Arctic Council', *American Journal of International Law*, vol. 93, no. 3 (1999).
¹¹ Arctic Council, Rules of Procedure 1998, <http://www.arctic-council.org/index.php/en/about/documents/category/4-founding-documents>; and Scrivener, D., 'Arctic environmental cooperation in transition', *Polar Record*, vol. 35, no. 192 (1999).
¹² Arctic Council (note 11) and Scrivener (note 11).

attract the attention of the public. In the early years, civil servants at these meetings frequently represented member states. This is no longer the case; the 2011 ministerial meeting in Nuuk, for example, was attended by both Russian Foreign Minister Sergei Lavrov and US Secretary of State Hillary Clinton. The presence of the two countries' foreign ministers (Sergei Lavrov and John Kerry) added extra significance to the Kiruna meeting in 2013, and is indicative of the AC's increased importance in recent years both as a discussion forum and as a launch pad for binding agreements negotiated by the Arctic states. The declarations issued during the ministerial gatherings often reveal how the member states would like to see the AC evolve.[13] They give expression to the basic policy underlying the AC's work, and the participation of the Russian and US foreign ministers at the meetings gives these statements further political weight.[14]

At the official level, the SAOs convene at least twice a year. The SAOs serve as a liaison between ministerial and working group levels. They are senior civil servants empowered by their respective governments to manage and oversee the work of the AC on a daily basis.[15] However, most of the AC's activities are performed in the working groups, which identify and analyse the challenges faced by actors in the region, and develop the scientific knowledge base.[16] Such challenges include everything from rising mercury levels to guidelines for Arctic shipping. The working groups (see figure 6.1) have been described as the powerhouse of the AC.[17]

---

[13] Young, O. R., 'If an Arctic Ocean Treaty is not the solution, what is the alternative?', *Polar Record*, vol. 47, no. 4 (2011), p.333.

[14] Foreign Minister Lavrov did not attend the ministerial meeting in Iqaluit, Canada, in Apr. 2015. This was perhaps a reflection of the ongoing crisis in Ukraine. However, the reasons for his non-attendance are beyond the scope of the discussion in this chapter.

[15] Stokke, O. S., 'Regime interplay in Arctic shipping governance: explaining regional niche selection', *International Environmental Agreements: Politics, Law and Economics*, vol. 13, no. 1 (2013), p. 72.

[16] The Arctic Council has 6 working groups. For an overview see the Arctic Council's website at <http://www.arctic-council.org/index.php/en/about-us/working-groups>. See also Kankaanpää, P. and Young, O., 'The effectiveness of the Arctic Council', *Polar Research*, vol. 31 (2012).

[17] Stokke, O. S., 'En indre sirkel i Arktisk Råd?' [An inner circle in the Arctic Council?], *Nordlys*, 28 Apr. 2010 (in Norwegian).

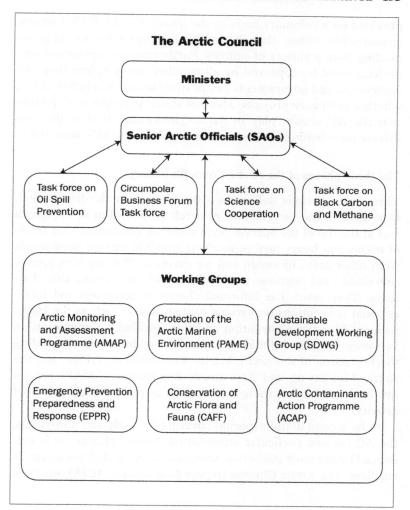

**Figure 6.1.** Arctic Council structure

*Note*: The Permanent Participants (six organizations representing Arctic indigenous peoples) take part at all levels.

It is worth noting that the AC does not have a separate programme budget, and funding for its activities and projects is provided on a voluntary basis by the member states.[18] This means, among other things, that the working groups need to apply for funding from a variety of sources. Furthermore, programmes and projects must be approved by all member states before they can commence, and governments can be quite selective when deciding whether to finance projects. Member states' perception of the role that the AC should play in Arctic governance is thus the real driving force behind (or in some cases against) the AC's activities.[19]

### The three phases of the Arctic Council's history

Three main periods define the AC's history. In the first period, from 1996 to the early 2000s, research on pollution in the Arctic was at the top of the agenda. Such research uncovered high levels of toxins and heavy metals, much of which is carried northwards from other areas by ocean and air currents. The working groups submitted (and continue to submit) reports addressing this challenge. Their work has informed climate negotiations and international conventions on various contaminants (such as the International Mercury Convention and the Stockholm Convention on Persistent Organic Pollutants).[20] Several working groups (including the Protection of the Arctic Marine Environment, PAME, and the Arctic Monitoring and Assessment Programme, AMAP) remain involved in work on these conventions as part of their principal activity.[21]

In the second defining period covering the early and mid-2000s, the AC focused particular attention on climate change, as is evidenced by the most publicized scientific work carried out under its auspices—the Arctic Climate Impact Assessment (ACIA), the final

[18] Stokke (note 15), p. 72.
[19] Rottem (note 3).
[20] Duyck, S., 'Which canary in the coalmine? The Arctic in the international climate change regime', eds T. Koivurova, G. Alfredsson and W. Hasanat, *The Yearbook of Polar Law*, vol. 4, 2012 (Brill: Leiden, 2012). Minamata Convention on Mercury (International Mercury Convention), opened for signature 10 Oct. 2013, not in force, <http://www.mercuryconvention.org/>. Stockholm Convention on Persistent Organic Pollutants, opened for signature 22 May 2001, entered into force 17 May 2004, 2256 *United Nations Treaty Series* 119, <https://treaties.un.org/pages/ViewDetails.aspx?src=TREATY&mtdsg_no=XXVII-15&chapter=27&lang=en>.
[21] See Kankaanpää and Young (note 16) for an analysis of the working groups' perception of their role and importance in international cooperation.

results of which were released in 2004.[22] According to the ACIA, the Arctic is in a unique position with regard to climate change: the first signs of the global consequences of climate change are being felt there. This period of the AC's history was thus marked by a clear focus on mapping the consequences of global warming and adaptation to climate change. The AC placed a number of challenging items on the mitigation agenda at that time, including limits on the emission of so-called short-lived greenhouse gases such as soot, methane and tropospheric ozone.

In recent years, in what can be said to be the third defining period, the AC has concerned itself with what can be done to adapt to climate change and to respond to the growing interest in the AC as a forum for international cooperation. This new phase is largely in response to the steady rise in activity in the North, partly as a result of the retreating sea ice cover. The SAR and the Oil Spill agreements are just two examples of the changing focus.

## International cooperation and participation

As noted above, the Arctic is influential in the foreign policy agendas of the Arctic states and others with a stated interest in the region.[23] All of the Arctic states have published Arctic strategy documents in recent years, for example.[24] Canada's 2009 Arctic foreign policy document sets out fairly explicitly Canada's views on the AC's future role. The AC, it says, is the key forum for collaboration on Arctic affairs and Canada will work to strengthen it.[25] Denmark shares this positive view of the AC. In its strategy

---

[22] Arctic Council, Arctic Climate Impact Assessment (ACIA) Secretariat, *Impacts of a Warming Arctic: Arctic Climate Impact Assessment* (Cambridge University Press: Cambridge, 2004). The idea for an assessment was first mooted in the mid-1990s, although work did not start until around 2000.

[23] Bailes, A. J. K. and Heininen, L., *Strategy Papers on the Arctic or the High North: A Comparative Study and Analysis*, Report series, Centre for Small State Studies, Institute of International Affairs, University of Iceland (University of Iceland: Reykjavik, 2012).

[24] For an overview of these strategies see the Arctic Council website at <http://www.arctic-council.org/index.php/en/document-archive/category/12-arctic-strategies>.

[25] Government of Canada, 'Statement on Canada's Arctic foreign policy: exercising sovereignty and promoting Canada's Northern Strategy abroad', 20 Aug. 2010, <http://www.international.gc.ca/arctic-arctique/assets/pdfs/canada_arctic_foreign_policy-eng.pdf>.

document Denmark states that it would welcome a change in the AC's remit so that it becomes a '"decision-making" organization'.[26]

The strategic documents of the three states without an Arctic coastline (Finland, Iceland and Sweden) also describe the AC as the central forum for addressing Arctic-related issues. They express their respective governments' commitment to strengthening the AC's role. However, key actors working on AC issues have claimed that non-littoral Arctic states have taken a less active part in the work of the AC (with the exception of Iceland).[27] This is perhaps understandable as the rights and responsibilities of the Arctic littoral states (Canada, the Kingdom of Denmark, Norway, Russia and the USA, the so-called Arctic Five) are much more comprehensive than those of non-littoral and non-Arctic states by virtue of the Arctic littoral state's sovereign rights and jurisdiction over large areas of the Arctic Ocean.

Both Russia and the USA, the AC's two great-power members, are also generally supportive of the AC's role. A US presidential directive on the Arctic issued in January 2009 highlights the importance of the AC in facilitating cooperation on Arctic issues.[28] This is also emphasized in the US 2013 Arctic strategy document.[29] The USA's positive attitude towards the SAR and Oil Spill agreements, moreover, could be seen as a concrete expression of support. In addition, the participation of Clinton and Kerry at the 2011 Nuuk meeting and the 2013 Kiruna meeting, respectively, is an indication of the significance of the Arctic in US foreign policy.[30]

The Russian Government has issued several documents detailing its Arctic strategy in the past few years. These stress the AC's

---

[26] Governments of Denmark, the Faroe Islands and Greenland, 'Kingdom of Denmark strategy for the Arctic 2011–2020', <http://um.dk/en/foreign-policy/the-arctic/>, p. 52.

[27] Rottem, 'A note on the Arctic Council agreements' (note 3).

[28] White House, National Security Presidential Directive (NSPD) 66 and Homeland Security Presidential Directive (HSPD) 25, 9 Jan. 2009, <http://georgewbush-whitehouse.archives.gov/news/releases/2009/01/20090112-3.html>.

[29] White House, 'National strategy for the Arctic region', 10 May 2013, <http://www.whitehouse.gov/sites/default/files/docs/nat_arctic_strategy.pdf>.

[30] For a detailed account of US involvement in the Council see Pedersen (note 4). It could be argued that the change from a Republican to a Democratic administration has caused a shift in US foreign policy in favour of the Arctic. The Macondo oil spill in the Gulf of Mexico in 2010 also raised the political stakes in the USA. For background see Rottem, 'A note on the Arctic Council agreements' (note 3).

influence in facilitating international cooperation.³¹ Russia also actively participated in the drafting of the SAR and Oil Spill agreements, and alongside other members spearheaded much of the preliminary work.

Norway, which is also a littoral state, chaired the AC from autumn 2006 to spring 2009, a key period in the formulation of Norway's High North policy and Arctic strategy. Norwegian interest in the AC rose at the same time, with the government describing the AC as 'the main multilateral forum in the North', highlighted by its efforts to establish a permanent secretariat in Tromsø.³² In a white paper published in November 2011 the Norwegian Government underlined its commitment to ensuring a well-functioning AC.³³ The document states that Norway's Arctic policy was developed mainly in the context of the AC.³⁴

*Differences of opinion*

Despite the overarching consensus on all sides as to the importance of the AC, disagreements between the Arctic littoral states and the three other member states have occurred. In 2008 an Arctic Ocean conference, organized by Denmark's Foreign Minister Per Stig Møller, took place in Ilulissat, Greenland. The main message from the conference was that the United Nations Convention on the Law of the Sea (UNCLOS) provides a solid framework for responsible management of the Arctic Ocean by the Arctic littoral states.³⁵ However, several other issues were also on the agenda, including polar oil and mineral exploration, maritime security, transportation and environmental regulations. The non-littoral Arctic states opposed any development that could have the effect of narrowing down the Arctic cooperation structure solely to the Arctic littoral states.

---

[31] Zysk, K., 'Russia's Arctic strategy: ambitions and constraints', *Joint Force Quarterly*, no. 57 (2010), pp. 102–10. Also see the discussion in chapter 4 of this volume.

[32] Government of Norway, 'Prop. 1 S (2009–2010)', Governmental proposal, 25 Sep. 2009, <https://www.regjeringen.no/no/dokumenter/prop-1-s-20092010/id581229/?ch=1&q=>, p. 106 (in Norwegian).

[33] Government of Norway, 'Meld. St. 7 (2011–2012): Nordområdene' [The High North—visions and strategies, White Paper no. 7 (2011–12)], 13 Feb. 2012, <https://www.regjeringen.no/no/dokumenter/meld-st-7-20112012/id663433/?ch=1&q=> (in Norwegian).

[34] Government of Norway (note 33), p. 78.

[35] United Nations Convention on the Law of the Sea, opened for signature 10 Dec. 1982, entered into force 16 Nov. 1994, <https://treaties.un.org/Pages/ViewDetailsIII.aspx?src=TREATY&mtdsg_no=XXI-6&chapter=21&Temp=mtdsg3&lang=en>.

Although the non-littoral Arctic states agreed with the main message from the conference that UNCLOS also applies to the Arctic Ocean, in their view, the AC was the appropriate venue for discussing all pan-Arctic challenges. The differences between the two sides no longer appear to be as divisive; nevertheless, they are still of importance when discussing the variable 'geometry of governance' over key issues, a topic that will be discussed later in this chapter.

*Observers and Permanent Participants*

At the 2013 Kiruna ministerial meeting, the question of observer status for the European Union (EU) and certain non-Arctic states headed the agenda. China, India, Italy, Japan, Singapore and South Korea secured such status at the meeting. However, the discussions over the EU were not as straightforward, with Canada's frosty relationship with the EU over the latter's ban on seal products thought to be one of the main stumbling blocks.[36] Although Canada and the EU seem to have found common ground, the EU has still not been granted permanent observer status. In addition, since the start of the crisis in Ukraine, which began towards the end of 2013, Russia appears to have hardened its resistance to the inclusion of the EU owing to Russia's deteriorating relationship with the EU and its member states. Thus, geopolitical tensions caused by the crisis in Ukraine are spilling over into Arctic cooperation.

At the ministerial meeting in Iqaluit, Canada, in April 2015 the ministers agreed to defer decisions on pending observer applications. Nevertheless, the desire for observer status on the part of the EU and several non-arctic states shows that the region is perceived as important by stakeholders outside the geographically limited Arctic region. However, this new-found interest in the AC by state and regional actors may have an impact on another important membership group—the Permanent Participants, which comprise organizations representing indigenous Arctic communities (see table 6.1).[37] Can the role of the observers be enhanced

---

[36] For an analysis of this topic see Wegge, N., 'Politics between science, law and sentiments: explaining the European Union's ban on trade with seal products', *Environmental Politics*, vol. 22, no. 2 (2013).

[37] For more information on the Permanent Participants see the Arctic Council's website at <http://www.arctic-council.org/index.php/en/about-us/permanent-participants>.

without undermining the position and influence of the Permanent Participants?

The involvement of the Arctic indigenous communities on issues of sustainable development and environmental protection has been and remains vital to the work of the AC. There are concerns that their visibility and influence in the AC would diminish were the number of observers to continue to rise. However, a key provision to be considered when reviewing an observer's contributions to the AC's activities is that such an observer should demonstrate political willingness to contribute to the work of the Permanent Participants and other Arctic indigenous groups.[38] How observers interpret and apply this provision is currently unclear and requires further debate.

It could be argued that it is in the best interest of the Arctic's indigenous people that non-Arctic maritime states strengthen their role in AC meetings and working group sessions. Observers would then be given the opportunity to work on Arctic issues in a cooperative and informed manner. If observers are not included in AC work they could turn to forums and processes where indigenous groups are marginalized. Moreover, the standards regulating observer participation preserve the formal privileges of the Permanent Participants. On the other hand, increased observer participation could extend the AC's agenda to such a degree that the Permanent Participants would need to prioritize certain issues and projects. Widening the agenda could also put pressure on capacity (both human and financial).

The Arctic has undergone a political renaissance over the past few years; the accent is now on cooperation rather than conflict.[39] In this context, the AC is perceived as being of relevance by all the Arctic states, Permanent Participants and many non-Arctic states. The Oil Spill Agreement, which is analysed in detail in the next section, provides an interesting recent example of the Arctic cooperation process in practice.

---

[38] Arctic Council, 'Observer manual for subsidiary bodies', adopted at the 8th Arctic Council ministerial meeting, Kiruna, 15 May 2013, <https://oaarchive.arctic-council.org/bitstream/handle/11374/939/2015-09-01_Observer_Manual_website_version.pdf?sequence=1&isAllowed=y>.

[39] The Ukraine crisis has consequences for Arctic cooperation; however, discussion of this topic is beyond the scope of this chapter.

**Table 6.1.** Arctic Council membership

| Arctic member states | Non-Arctic observer states | Permanent Participants |
|---|---|---|
| Canada | China | Aleut International Association (AIA) |
| Kingdom of Denmark | France | Arctic Athabaskan Council (AAC) |
| Finland | Germany | Gwich'in Council International (GCI) |
| Iceland | India | Inuit Circumpolar Council (ICC) |
| Norway | Italy | Russian Association of Indigenous Peoples of the North (RAIPON) |
| Russia | Japan | Saami Council (SC) |
| Sweden | South Korea | |
| United States/Alaska | The Netherlands | |
| | Poland | |
| | Singapore | |
| | Spain | |
| | United Kingdom | |

Source: The Arctic Council's website, <http://www.arctic-council.org/index.php/en/about-us/permanent-participants>.

The following review discusses the nature of the agreement, the potential for the AC to take on even more of a decision-making role, and how the AC fits within the framework of international and national governance in the Arctic.

## III. The Oil Spill Agreement

### Identifying the need for an agreement

The Oil Spill Agreement was signed at the 2013 ministerial meeting in Kiruna and is structured along the same lines as the SAR Agreement signed in 2011. It defines the respective states' areas of responsibility and emphasizes the need for cooperation. The exploration, extraction and transport of oil and gas are three of the main threats to the Arctic environment at present.[40] According to the AMAP working group in the AC, as the rate of such activity grows, so too will the size of the environmental protection challenge.[41] Recognition of this fact was central to the 2004 ACIA, and concern over the potential for oil spills also spurred the AC's working groups to prepare non-binding guidelines for oil and gas operations in the region.[42]

Undiscovered oil and gas deposits have led to a growth in interest in the Arctic region. In 2000 the US Geological Survey (USGS) presented figures indicating that 24 per cent of the world's undiscovered oil and gas resources might be found in the Arctic.[43] The USGS issued a new report in 2008 stating that 13 per cent of undiscovered global oil resources and 31 per cent of undiscovered global gas resources are located in the Arctic.[44] Nonetheless, despite the considerable attention given to Arctic offshore petroleum resources in recent years, actual industrial activity remains very limited. Only two fields are in production on the Arctic continental shelf—one in Norway and one in Russia. The main activity is exploration and there are significant differences in the level and organization of offshore petroleum activity in the various Arctic littoral states.

In Canada, Greenland and the USA the initiative is clearly in private hands, and only in Greenland is there seemingly strong public support for increased activity. In Norway and Russia the state is more directly involved through majority ownership of the dominant companies and the setting of development priorities. However, each major investment project has its own characteristics, and the

---

[40] Arctic Council, Arctic Monitoring and Assessment Programme (AMAP), *AMAP Assessment Report: Arctic Pollution Issues* (AMAP: Oslo, June 1998); and Arctic Council, Protection of the Arctic Marine Environment (PAME), 'Arctic offshore oil and gas guidelines', 10 Oct. 2002, <http://www.pame.is/images/03_Projects/Offshore_Oil_and_Gas/Offshore_Oil_and_Gas/ArcticGuidelines.pdf>.

[41] Arctic Council, Arctic Monitoring and Assessment Programme (AMAP), *Arctic Oil and Gas 2007* (AMAP: Oslo, 2007).

[42] Arctic Council, Protection of the Arctic Marine Environment (PAME) (note 40).

[43] US Geological Survey (USGS), 'World petroleum assessment 2000', June 2000, <http://energy.usgs.gov/OilGas/AssessmentsData/WorldPetroleumAssessment.aspx#3882218-research>.

[44] US Geological Survey (USGS), 'Circum-Arctic resource appraisal: estimates of undiscovered oil and gas north of the Arctic Circle', USGS Fact Sheet 2008-3049 (2008), <http://pubs.usgs.gov/fs/2008/3049/fs2008-3049.pdf>.

speed and level of Arctic offshore petroleum development both seem to have reduced in recent years. This is especially true for Alaska in the USA, but there are general concerns over costs in other parts of the Arctic too—partly caused by the increased focus on environmental protection. In addition, the revolution in the natural gas market caused by the boom in shale gas has made Arctic offshore gas much less commercially attractive. Despite such changes in the market, work on oil spill prevention, preparedness and response is needed in the region.

The growth in shipping activity is also relevant when assessing potential oil spill scenarios in Arctic waters. There is wide variation in the type and location of shipping traffic when looking at the region as a whole. For example, in 2013, 71 ships transited the Northern Sea Route along the Russian Arctic coast, a route Russia hopes to establish as its northern export highway.[45] However, the 2014 shipping season saw a dramatic drop in tonnage. Thus, one should be critical of any large expectation towards Arctic shipping in the near future. There will be great annual and seasonal variations. Ice conditions in the Northwest Passage are even harder, and a significant increase in commercial shipping is unlikely there. However, interest in North American Arctic development and tourism is growing, which may lead to greater cruise traffic in Canadian and US Arctic waters.[46]

Cruise traffic is increasing in Greenlandic waters too, resulting in discussions as to the capacity of coast guards and accident and rescue teams to deal with a major accident. In 2012 and 2013 the Kingdom of Denmark conducted two search and rescue exercises some distance from Greenland's east coast. In both cases the scenario was that a cruise ship had run into difficulties in a remote Arctic area. The potential for an oil spill hazard was also relevant in this context.[47] Both exercises were evaluated, and challenges were revealed regarding the infrastructure and capacity to cover

---

[45] Humpert, M., *Arctic Shipping: An Analysis of the 2013 Northern Sea Route Season* (Arctic Institute: Washington, Oct. 2013).

[46] Office of the Auditor General of Canada, *Report of the Commissioner of the Environment and Sustainable Development: Chapter 3: Marine Navigation in the Canadian Arctic* (Office of the Auditor General of Canada: Ottawa, 2014).

[47] Armed Forces of Denmark (Forsvaret), Final report from search and rescue exercise Greenland Sea 2013 (SAREX 2013), 1 Nov. 2013, <http://www2.forsvaret.dk/viden-om/organisation/arktisk/SAREX/Pages/SAREX.aspx>.

the vast distances in the Arctic. The evaluations also noted the increased likelihood of extreme weather events caused by a changing climate and possibly leading to an increase in accidents in the region.

What the above examples have in common is the light they shed on the capacity and infrastructure challenges faced by those operating in the Arctic. This insight is of particular relevance to Norwegian Arctic waters as 80 per cent of Arctic shipping passes through those waters.[48] The start of and rise in hydrocarbon exploration and development, Norway's large fishing fleet, and increased tourism (especially around the Svalbard archipelago) are putting added pressure on the existing Arctic infrastructure. Such developments have moved the question of international and regional cooperation on maritime safety on to the international political agenda.

As already mentioned, the AC has been exploring the political and scientific ramifications of oil and gas extraction, and the likely growth in this field. This work has been undertaken primarily by the working groups, with Norway assuming an active role in the process. As early as 2009 the Norwegian Government suggested that AC member states should negotiate an oil spill agreement. There are several possible explanations for Norway's firm backing of the agreement and its co-chairmanship of the task force appointed to draw it up. It could be argued, for example, that Norway considered the agreement to be a way of enhancing the legitimacy of its oil and gas industry in the North.[49] In other words, there could be an element of pre-emption in the drafting of the agreement: such an agreement could be viewed as constituting indirect acceptance by the AC member states of oil and gas operations in an area that some critics would like to see closed to the industry altogether.

---

[48] Government of Norway, 'Norway's arctic policy', 10 Nov. 2014, <https://www.regjeringen.no/contentassets/23843eabac77454283b0769876148950/nordkloden_rapport-red.pdf>.

[49] Rottem, 'A note on the Arctic Council agreements' (note 3).

## The relevance and limits of the agreement

In 2011 a task force was appointed to draft the text of the Oil Spill Agreement. It was co-chaired by Norway, Russia and the USA. The agreement was signed in May 2013. Article 1 sets out the main objective of the agreement, which is 'to strengthen cooperation, coordination and mutual assistance among the Parties on oil pollution preparedness and response in the Arctic in order to protect the marine environment from pollution by oil'.[50] The agreement refers to several other international obligations, each of which provides a framework for oil spill preparedness and response, the most pertinent being the 1990 International Convention on Oil Pollution Preparedness, Response and Cooperation, and the 1969 International Convention Relating to Intervention on the High Seas in Cases of Oil Pollution Casualties.[51] Note is made moreover of the work done under the auspices of the International Maritime Organization (IMO), a specialized agency of the UN and UNCLOS. While the Oil Spill Agreement is thus part of a broader regime, it is nonetheless the first regional agreement on oil spill preparedness in the Arctic.[52]

The agreement consists of 23 articles and 5 appendices specifying, among other things, the framework of national contact points, and several non-binding operational guidelines. The latter have 'provisions to guide cooperation, coordination and mutual assistance for oil pollution preparedness and response in the Arctic' (Appendix IV). Two important general factors stand out

---

[50] Agreement on Cooperation on Marine Oil Pollution, Preparedness and Response in the Arctic (note 2).

[51] International Convention on Oil Pollution Preparedness, Response and Cooperation, opened for signature 30 Nov. 1990, entered into force 13 May 1995, 1891 *United Nations Treaty Series* 77, <https://treaties.un.org/doc/Publication/UNTS/Volume%201891/volume-1891-I-32194-English.pdf>. International Convention Relating to the Intervention on the High Seas in Cases of Oil Pollution Casualties, opened for signature 29 Nov. 1969, entered into force 6 May 1975, 970 *United Nations Treaty Series* 211, <https://treaties.un.org/doc/Publication/UNTS/Volume%20970/volume-970-I-14049-English.pdf>.

[52] Vinogradov, S., 'The impact of the deepwater horizon: the evolving international legal regime for offshore accidental pollution prevention, preparedness and response', *Ocean Development and International Law*, vol. 44, no. 4 (2013), p. 351.

during an analysis of the relevance of the agreement, namely capacity and organization. Both aspects are touched on below.[53]

Article 6 expresses the commitment of the parties in the event of an accident to notify other states likely to be affected. How such a system would work in practice, however, is difficult to assess. The agreement does not detail the resources needed for such operations beyond the minimum national system in place (i.e. Article 4 refers to 'national contingency plan or plans for preparedness and response to oil pollution incidents').

The agreement repeatedly urges the signatories to promote cooperation and information exchange. This will be partly reflected in joint exercises in which other stakeholders can take part and where the relevant provisions of the agreement can be implemented in the field (Article 13(3) and (4)). So far, one exercise has been conducted in Canada, and one is being planned by the USA during its AC chairmanship.[54] In addition, the agreement states that 'each Party shall bear its own costs deriving from its implementation of this Agreement' (Article 15(1)). The agreement does not enhance the financial capacity of the AC to deal with oil spills, nor does it impose any obligation on parties to increase the resources devoted to oil recovery. The absence of such provisions has drawn criticism from, among others, non-governmental organizations such as Greenpeace and the World Wildlife Fund for Nature (WWF).[55] Critics also urged the parties to ban heavy oil in the region and to address the challenges arising from the lack of technology to clean up oil in ice-covered waters.[56]

No new formal structures beyond the Meetings of the Parties are provided for under the agreement (Article 14). The agreement encourages the signatories to hold the first such meeting no later

---

[53] This section of the chapter draws on Rottem, 'A note on the Arctic Council agreements' (note 3).

[54] For further information on the exercise conducted in Canada see Arctic Council, 'Arctic exercise: after action report on the Agreement on Cooperation on Marine Oil Pollution Preparedness and Response in the Arctic', Sep. 2014, <https://oaarchive.arctic-council.org/handle/11374/404>. Information on the proposed exercise during the US chairmanship provided in communications with the author from a member of Arctic Council, Emergency Prevention Preparedness and Response (EPPR) working group, 15 Mar. 2015.

[55] The WWF was represented in the Norwegian delegation drafting the agreement and enjoys observer status in the Arctic Council.

[56] Rottem, 'A note on the Arctic Council agreements' (note 3).

than one year after the agreement's entry into force. Among other things, the meetings will serve as an opportunity to discuss and agree amendments to the appendices and can be held in conjunction with general AC meetings. Moreover, bodies involved in operational issues, but that are not necessarily part of the AC, may participate in discussions on matters of importance to the agreement (Article 14(2)). While no new organizational structures have been put in place to oversee the implementation of the agreement, it could be argued that the Emergency Prevention, Preparedness and Response Working Group (EPPR) under the AC could perform this function. Moreover, the EPPR is responsible for keeping the agreement's Operational Guidelines updated. But, as at the time of writing, it is the signatories themselves that must take steps to facilitate and monitor progress. Thus, the states chairing the AC have a crucial role to play in this regard.

Although the agreement may help to develop cooperation between Arctic states, it remains arguably of secondary importance in the current context. A number of parallel and complementary mechanisms to promote cooperation already exist outside the agreement. The collaboration between Norway and Russia is particularly notable and is perceived by key stakeholders as being successful.[57] Norway and Russia signed an agreement on oil spill response in the Arctic as early as 1994, thus providing for a joint, integrated contingency plan for oil pollution in the Barents region, and establishing, among other things, guidelines on notification procedures and joint exercises.[58] Other bilateral cooperation mechanisms have subsequently been added to the agreement between Norway and Russia.[59]

It is also the case that nationally formulated preparedness and safety guidelines imposed on the industry tend to dictate how oil and gas exploration should be conducted in the Arctic. It is only in extreme situations that a government will itself lead an oil spill

---

[57] Sydnes, A. and Sydnes, M., 'Norwegian–Russian cooperation on oil-spill response in the Barents Sea', *Marine Policy*, vol. 39, no. 1 (2013), p. 260.

[58] Norwegian Ministry of Foreign Affairs, 'Opportunities and challenges in the North', Report no. 30 (2004–2005) to the Storting, 15 May 2005, <https://www.regjeringen.no/globalassets/upload/kilde/ud/stm/20042005/0001/ddd/pdts/stm200420050001ud_ddd pdts.pdf>.

[59] Sydnes and Sydnes (note 57), p. 260.

response. The critical task for each government is therefore to set standards for the industry, and the Arctic states currently have differing regulatory regimes for oil and gas extraction.[60]

With regard to shipping, work done under the IMO (e.g. the drafting of a Polar Code, discussed below) is central, but in terms of oil spill preparedness and response in particular, national regulations and capacities are key. Thus, governance in this area could be categorized as a regulatory regime with different institutions performing complementary functions—a point that will be discussed in the next section of this chapter.[61] The main purpose of the Oil Spill Agreement is not to set standards or specify capacity levels, but to coordinate collaboration through the exchange of information and conduct of joint exercises.

Furthermore, the agreement emphasizes the primacy of UNCLOS as the leading international framework. The preamble to the agreement refers to the need to take relevant provisions of UNCLOS into consideration where necessary. Article 16 is even more specific: 'Nothing in this Agreement shall be construed as altering the rights or obligations of any Party under other relevant international agreements or customary international law as reflected in the 1982 United Nations Convention on the Law of the Sea [UNCLOS]'. In a Norwegian context, it is a perception that has defined the development of the government's Arctic policy. This basic standard is immutable.

What the above analysis shows is that it is difficult to identify any operational consequences of the agreement (apart from the exercise conducted in Canada). While it is perhaps too early to make any clear conclusions as to the impact of the agreement, it has been criticized for an alleged lack of ambition.[62] The agreement has more importance as a symbol of Arctic cooperation, then, than as a practical mechanism. The wording of the agreement places few obligations on the signatories, and more sophisticated regional agreements, such as those between Norway and Russia, will continue to be of greater relevance. The value of the agreement should

---

[60] Dagg, J. et al., *Comparing the Offshore Drilling Regulatory Regimes of the Canadian Arctic: The US, the UK, Greenland and Norway* (Pembina Institute: Drayton Valley, June 2011).

[61] Stokke (note 15).

[62] Rottem, 'A note on the Arctic Council agreements' (note 3).

not be dismissed completely, however, since it forms part of a larger regulatory universe. It can be used to establish and underpin formal and informal interorganizational action, and streamline communication procedures in the event of a major oil spill in the Arctic. Nevertheless, in the case of Norway, for example, bilateral cooperation with Russia will remain of prime importance, for geographical and institutional reasons.[63]

## IV. The Arctic Council in Arctic governance

The Oil Spill Agreement is suited to the framework of the AC. The AC has spent a great deal of time and effort over the years dealing with matters of relevance to the agreement. The working groups have accumulated unique expertise through their regular assessment of the environmental measures necessary to respond to the growth in commercial activity in Arctic waters. However, as mentioned previously, this is an agreement centred on information exchange and joint exercises, and the focus is on preparedness and response not preventive measures. The agreement is limited in scope and has no guidance regarding oil and gas operations in the region, for example, because the Arctic littoral states are wary of limiting their sovereignty in an area of strategic importance.[64] Thus, it can be argued that the agreement will have a greater effect only once the standard of conduct for the oil and gas sector (and shipping and tourism industries) has been defined and implemented by national governments and in international forums. The AC's role in Arctic governance currently falls somewhere in between the international and national frameworks.

### The Arctic governance framework

*Legally binding mechanisms*

As noted earlier, UNCLOS provides the fundamental international legal framework for governance in Arctic waters. By ensuring

---

[63] Sydnes, M. and Sydnes, A., 'Oil spill emergency response in Norway: coordinating interorganizational complexity', *Polar Geography*, vol. 34, no. 4 (2011).

[64] Stokke, O. S., 'Environmental security in the Arctic: the case for multilevel governance', *International Journal*, vol. 66, no. 4 (2011), p. 847.

rights and responsibilities of coastal states and flag states, and providing the basis for delimiting maritime zones according to their legal status, UNCLOS remains the core international legal framework for all human activities at sea. As to continental-shelf resources, littoral states enjoy exclusive management authority but they are, however, strongly encouraged 'to harmonize their policies in this connection at the appropriate regional level' (Article 208 of UNCLOS). As UNCLOS grants littoral states jurisdiction within their exclusive economic zones, such states have the right and responsibility to respond to incidents threatening the marine environment. Thus, the predominant mode of governance for Arctic petroleum activities will continue to be unilateral management by each of the Arctic littoral states, and UNCLOS grants littoral states relatively free control in regulation of continental-shelf activities, a point that was firmly underlined by the Arctic littoral states at the meeting in Ilulissat in 2008.

Nonetheless, several international and transnational norm-making processes impact on Arctic littoral-state regulation and govern petroleum-related activities, particularly in relation to shipping. The community of stakeholders in this area is diverse and wide ranging, and government agencies have key roles in the supervision and formulation of standards and regulations.[65]

The maritime transport activities necessary for exploration, development, and production of hydrocarbons are mostly subject to flag state jurisdiction, so effective regulation requires global action under the IMO. The IMO has emerged as the primary arena for crafting governance arrangements for Arctic shipping. The IMO activities most relevant to Arctic oil and gas concern platform-related provisions of the International Convention for the Prevention of Pollution from Ships (MARPOL Convention) and the development of a mandatory Polar Code for vessels that operate in ice-covered waters.[66] All the Arctic states are parties to the

---

[65] For an introduction to the Norwegian regulations see the fact sheet issued by the Norwegian Petroleum Directorate in 2013: Norwegian Petroleum Directorate, 'Facts 2013: the Norwegian petroleum sector', Mar. 2013, <http://npd.no/en/Publications/Facts/Facts-2013/>. For a pan-Arctic analysis see Dagg et al. (note 60).

[66] The International Convention for the Prevention of Pollution from Ships (MARPOL Convention) adopted 2 Nov. 1973, entered into force 2 Oct. 1983. The Polar Code is expected to enter into force on 1 Jan. 2017.

MARPOL Convention, which places legally binding restrictions on emissions and discharges. The IMO Polar Code negotiations aim at strengthening the substance, scope and form of the 2002 Guidelines for Ships Operating in Arctic Ice-covered Waters to generate more stringent and legally binding requirements concerning vessel construction and equipment, training and discharges.[67] This work responds to the special challenges that exist in Polar areas, such as icing, poor satellite coverage and hydrography, and limited emergency response capacity. Accordingly, work under the auspices of the IMO, including guidelines and expertise of navigation in the area, along with national measures (such as the creation of shipping lanes) will have the greatest preventive effect in terms of oil spill hazards.[68]

Additionally, some segments of the Arctic shelves are subject to mandatory rules developed under the Convention for the Protection of the Marine Environment of the North-East Atlantic (OSPAR Convention).[69] Among the Arctic states, the Kingdom of Denmark and Norway are bound by these rules, as are 13 non-Arctic littoral states and the EU.[70] The OSPAR Convention prohibits the disposal and abandonment of any offshore installation at sea, with certain exceptions subject to a national decommissioning permit. Moreover, the OSPAR offshore oil and gas strategy sets out discharge regulations that are more stringent than those globally applicable under the IMO, especially with respect to chemicals and oil in produced water.[71]

---

[67] International Maritime Organization, 'Guidelines for ships operating in Arctic ice-covered waters 2002', MSC/Circ.1056, MEPC/Circ.399, 23 Dec. 2002.

[68] The work done under the auspices of the Association of Arctic Expedition Cruise Operators (AECO) is also relevant in this regard. For further details see the AECO's website at <http://www.aeco.no/>.

[69] The OSPAR Commission, named after the original Oslo and Paris conventions, is the mechanism by which 15 governments of the western coasts and catchments of Europe, together with the European Union, cooperate to protect the marine environment of the North-East Atlantic. The Convention for the Protection of the marine Environment of the North-East Atlantic (OSPAR Convention), opened for signature 22 Sep. 1992, entered into force 25 Mar. 1998.

[70] The 15 governments party to the convention are Belgium, the Kingdom of Denmark, Finland, France, Germany, Iceland, Ireland, Luxembourg, the Netherlands, Norway, Portugal, Spain, Sweden, Switzerland and the UK.

[71] Further details of the OSPAR Commission's 'North-East Atlantic environment strategy', including the 'Offshore industry strategy', are available at its website: <http://www.ospar.org>.

## Non-binding mechanisms

With regard to sea-shelf activities, the Arctic littoral states have also committed themselves to a certain level of oversight by several 'soft-law' institutions, and it is within this group that the AC has relevance. The AC is listed in the *Yearbook of International Organizations 2012–13* as a 'limited or regionally defined' organization with intergovernmental and international organizations as members.[72] However, the AC lacks the ability to make binding decisions, meaning that mechanisms such as the Oil Spill Agreement are negotiated under the auspices of the AC between the eight Arctic states. While not definitive criteria, international organizations that can take binding actions and therefore provide governance direction generally possess three characteristics: (*a*) a hard-law instrument of formation, (*b*) at least one subordinate unit (or an organ) that can operate independently, and (*c*) establishment and recognition under international law. The AC cannot operate separately from the eight Arctic states that created it, nor can it obligate other states or organizations to take specific measures because of its soft-law nature. The AC, therefore, lacks international legal personality.

However, since 2002 the AC has upheld a set of Arctic Offshore Oil and Gas Guidelines and numerous other soft-law instruments that summarize best environmental practices, and—as discussed previously—it has provided a forum for negotiating a legally binding agreement on marine oil pollution preparedness and response.[73] However, as noted earlier in this chapter the ambition of the Oil Spill Agreement should not be exaggerated. It aims to improve the coordination of Arctic littoral-state capabilities and does not seek to limit the exercise of their sovereign rights in the management of their shelf activities, nor does it address capacity growth. Tellingly, the mandate of a task force set up by the AC in 2013 to prepare an instrument regarding the prevention of oil spills made no mention of legal commitments.[74] Nonetheless, the work

---

[72] Union of International Organizations, *Yearbook of International Organizations 2012–13* (Brill: Leiden, 2012).

[73] Arctic Council, 'Arctic offshore oil and gas guidelines' (note 40).

[74] Stokke, O. S., 'The promise of involvement: Asia in the Arctic', *Strategic Analysis*, vol. 37, no. 4 (2013), pp. 474–79.

done under the auspices of the AC has played a role in energizing negotiations within the IMO, and may therefore have had some influence in the creation of preventive measures.

Alongside the binding or non-binding governmental processes discussed in this section, industry-based private governance is increasingly significant for Arctic petroleum activities. A major accident involving a large oil spill in the Arctic would have severe repercussions on the entire scope of industries involved in regional hydrocarbon resource development. Thus, through arrangements like the Barents 2020 initiative companies involved in Arctic hydrocarbon extraction, such as Gazprom, Rosneft, Statoil and Total, have developed standards for oil and gas operations in the Barents Sea that have influenced the International Organization for Standardization's (ISO) work on cold region petroleum and natural gas activities.[75]

### The future role of the Arctic Council

The above discussion shows that legally binding commitments have become more stringent over time, and Arctic littoral states, as well as flag states and industry actors, are also increasingly observant of non-binding or privately developed norms regarding commercial operations on the Arctic shelves. However, the constraints that international and transnational regimes place on the Arctic littoral states with respect to offshore oil and gas activities are relatively loose and go no further than each state has been prepared to accept. This is also very much apparent in the work of the AC and in the Oil Spill Agreement, in particular. What does this mean for the AC's potential to become a platform for the creation of future agreements of significance to developments in the Arctic? In recent debates three issues have received attention: fisheries, biodiversity/protected areas and mitigation measures (short-lived climate forcers).[76] However, on these issues the AC has again shown itself lacking as a platform for deeper cooperation.[77]

---

[75] International Organization for Standardization (ISO), ISO 19906:2010 Petroleum and natural gas industries—Arctic offshore structures, <http://www.iso.org/iso/catalogue_detail.htm?csnumber=33690>.
[76] Rottem, 'A note on the Arctic Council agreements' (note 3).
[77] Stokke (note 64) p. 836.

## Arctic fishing regulation

The management of Arctic fishing industry resources is based on migration patterns across the countries' various economic zones. Regional and often bilateral arrangements and arenas (such as the Norwegian–Russian Fisheries Commission) are thus the more natural choice for forging agreements. However, if the international fishing fleet were to move deeper into the Arctic Ocean new regulatory mechanisms might be needed. Yet great uncertainty surrounds the likelihood of the discovery of new and commercially attractive fisheries in Arctic waters. The Arctic littoral states have taken steps to establish precautionary principles (stating that more research is needed) with regard to fishing in the Arctic Ocean.[78] The main challenge is that a large part of the central Arctic Ocean is high seas and thus open to access for several countries. The Arctic littoral states lack the authority to make decisions about such matters without reaching out to non-arctic states with an interest in fisheries in the Arctic Ocean. Thus, the Arctic littoral states (and Iceland) recognize the necessity of consulting non-arctic states. The AC is therefore not the most appropriate venue for expanding cooperation (resulting ultimately in a binding international agreement) on fisheries in the Arctic Ocean.

## Arctic biodiversity and protected areas regulation

In relation to issues of biodiversity and protected areas, the biggest challenge is again one of sovereign rights and jurisdiction. Understandably, the Arctic littoral states want control over their respective economic zones and continental-shelf areas. This is also evident in the strategies adopted for the region by the majority of Arctic states. Only Finland and Sweden (not Arctic littoral states) are willing to contemplate any regulation of the Arctic in this area under the auspices of the AC. Thus, the AC's role would most likely be as a provider of information on which to base decision making, rather than as a forum for the negotiation of an agreement between the eight member states.

---

[78] See e.g. Kramer, E., 'Accord would regulate fishing in Arctic waters', *New York Times*, 16 Apr. 2013.

*Climate change mitigation measures*

In the early and mid-2000s climate issues received heightened attention from the AC in its work, including a stronger focus on mitigation measures. In 2009 a task force was appointed to investigate short-lived climate forcers. This could have led to a recommendation to negotiate an internationally binding agreement under the auspices of the AC. However, the creation of the task force has so far not resulted in such a recommendation. While there are several reasons for this, the main argument is that although the release, for example, of black carbon in the polar regions would be expected to have a greater impact on the climate in polar regions than emissions in non-polar regions, implementing mitigation measures is essentially a global issue. It should therefore be addressed in other international forums. Again, the AC's main role would be as a producer of knowledge to feed into processes with wider international participation.

# V. Concluding remarks

The Arctic Council's significance and influence is growing. Arctic littoral states are mostly sympathetic to the idea of expanding the AC and giving it a more prominent role. As discussed in this chapter, the positivity surrounding the AC has enabled the creation of a permanent secretariat at Tromsø, and the signing of two binding international agreements negotiated by the eight permanent members. However, in light of climate change and the growth in commercial operations in the Arctic, as well as the general perception of the AC as a key forum for cooperation in the region, it is perhaps not so surprising that the member states have signed agreements on search and rescue, and oil spill preparedness and response.

From the Norwegian perspective, the analysis presented in this chapter has shown that the Oil Spill Agreement, in particular, has had little practical impact on Norwegian policy making in the Arctic in terms of expanding capacity and changing organizational structures. While the agreement's effect is limited, the common resources, joint exercises and shared experiences it envisages are not without importance. The agreement also has some political and symbolic value to the AC. It is, nevertheless, too early to say whether the Oil Spill Agreement will gain the substantive force some would like it to have.

It is difficult (and perhaps not particularly useful) to imagine a further expansion of the AC's operational scope. The AC will remain to all intents and purposes a decision-shaping body rather than a decision-making one. This does not mean that the AC will have no decisive impact on Arctic governance in the years ahead. States both with an Arctic coastline and without are represented by the AC, and its value as a forum for consulting and discussing Arctic issues with indigenous groups (the Permanent Participants) must be underlined. The AC is a convenient and appropriate venue for the creation of some aspects of Arctic policy, in close cooperation with the observers and Permanent Participants. Of equal importance is its position as a producer of knowledge within the wider patchwork of international bodies whose work affects the Arctic. The activities undertaken under the leadership of the AC help set the Arctic agenda. How much influence agreements negotiated under its auspices exert on the everyday activity in the Arctic is, however, as this chapter shows, a more complex question to answer.

# 7. Conclusions

LINDA JAKOBSON AND NEIL MELVIN

## I. In summary

Regional and international interest in the Arctic has grown substantially over the last two decades. This interest has been driven primarily by the challenges of climate change and, as the Arctic becomes more accessible, by the prospect of new opportunities to exploit the region's shipping routes and abundant resources. Despite military competition and even occasional confrontation, the relatively benign regional environment of the 1990s and 2000s proved ideal for the emergence of new forms of international cooperative governance in the Arctic. Indeed, for much of this period the region appeared to embody the idealist spirit of the immediate post-cold war era, bringing East and West together to try to deal with environmental problems and to build shared prosperity.

Since the eight Arctic states signed the Arctic Environmental Protection Strategy in 1991, a new architecture of formal governance has been created in the region, focused on the Arctic Council (AC). Established in 1996 through the Ottawa Declaration, the AC had humble beginnings as a principally intergovernmental body intended to promote scientific cooperation. Yet it has emerged as a political body that has a degree of influence and functions as a platform for debating the Arctic's key issues, as well as a forum for building consensus and forging new agreements.

This report is concerned with the key themes and questions relating to Arctic governance. In this concluding chapter, some of the major reflections contained in the various chapters are summarized, and this is followed by a clustering of some of the key issues that are likely to impact on the future of Arctic governance.

### Defining security in the Arctic

In chapter 2, Alyson Bailes addressed the issue of defining security in the Arctic. She noted that the Arctic is not a discrete security space but rather is co-opted into a variety of other security spaces. For much of the cold war, the region was one of the key strategic zones in the struggle for nuclear superiority between the superpowers. It was this struggle that promoted a militarization of the Arctic and a focus on hard security. The post-cold war experience has been remarkably different: while the Arctic has retained its strategic position for nuclear rivalry, notably between Russia and the United States, the hard security dimension of the region has declined significantly.

This transformation of the significance of the region for traditional security concerns has seen a broadening of the understanding of security. The Arctic has also opened up for new economic activities due to climate change and the Arctic states have increasingly sought to consolidate their sovereignty in the region, including through strengthening 'sovereignty supporting' security activities, such as policing and search and rescue (SAR). Thus the region's growing transnational linkages have further stretched the concepts of security in new directions.

As a result, Arctic security is seen to involve an increasingly complex agenda. Four distinct but overlapping dimensions are identified by Bailes: (*a*) hard security; (*b*) environmental, energy and economic issues; (*c*) managing civil emergencies; and (*d*) societal and human security. As a reflection of this comprehensive approach to regional security issues, the concept of a security actor has also been broadened in key areas—beyond the traditional focus on the state. While it could be said that the Arctic is characterized by weak cohesion and lacks a security community, collective approaches do exist and are quite effective in several of the nonmilitary dimensions of security (e.g. shipping safety, environmental standards, and SAR, as well as oil spill agreements).

Yet despite a broader understanding both of security and security actors, Bailes noted that sovereignty has remained the central ordering principle of security in the region. The eight Arctic states have ensured that security management and governance reflects their primacy, even while allowing other actors into the region. The most recent expansion of the AC permanent observers included key non-Arctic countries, notably China. Thus, the Arctic has increasingly become a model of 'messy' governance: not so much because of go-it-alone national approaches, but more because areas of national jurisdiction coexist with binding global regimes and emergent, sometimes equally binding, regional ones.

The reality of sovereignty as critically defining security in the region stands in the way of the emergence of a circumpolar security space. Instead, the Arctic remains essentially a set of subregions and, as such, Arctic security is susceptible to developments elsewhere. While the first two post-Soviet decades were conducive to an Arctic 'spirit of cooperation' (particularly between Russia and the transatlantic community), the crisis in Ukraine has raised the question of whether the basics of Arctic security will, once again, be changed.

## Security perceptions within the subregions

In chapter 3, Kristofer Bergh and Ekaterina Klimenko charted the perceptions that inform approaches to Arctic security within the Arctic subregions: North America (Canada and the USA), Europe (the Kingdom of Denmark, Finland, Norway and Sweden) and Russia. The diverse perspectives confirmed Alyson Bailes' conclusion that there is little prospect of the emergence of a 'genuine security community' in the Arctic in the short term.

Bergh and Klimenko noted that the contemporary institutions of Arctic governance have resulted from a shift in perceptions about regional security: from seeing the Arctic as a potential area of security threats to seeing the provision of security as a key to unlocking opportunities in the region. Thus, while science and environmental issues have driven the creation and consolidation of the AC, these developments have been underpinned by a new security context in the region.

For all Arctic states, perceptions of the region are infused with understandings of their interests. Therefore, the key question of security has been focused on how security policies can consolidate sovereignty, as the region opens up due to climate change and new interests in resource extraction and transportation. Despite significant differences in perception between the three main Arctic subregions—and even frictions and disputes within particular subregions, for example, between Canada and the USA over the North West Passage—a balance has emerged. This balance pivots on the idea of cooperation being the best way to advance national sovereignty and security in the Arctic.

Yet harder security developments have not been absent from the region. As Bergh and Klimenko noted, with the exception of strategic nuclear issues, security issues have been co-opted into the central discourse on cooperation surrounding the opening of the Arctic and protection of the region's environment. While the issues associated with the development and protection of the Arctic (soft security) have been addressed through the AC, harder security issues have been dealt with through informal multilateral meetings of security officials and bilateral cooperation, for example, regarding military training.

Underpinning the broadly cooperative security approach to the Arctic has been the relatively benign relationships that the Arctic states have had with each other elsewhere. Thus, as relations between Russia and Europe and the USA have become more fractious in Eastern Europe and the Middle East, the Arctic has been drawn into the wider dispute. In the light of these developments, can this cooperative Arctic spirit continue? The Ukraine crisis, in particular, has had a major impact on the core relationships of the Arctic states and may eventually lead to security developments in the region becoming part of the wider confrontation between Russia and the transatlantic community.

### Russian policy

In chapter 4, Andrei Zagorski outlined Russia's vital role in the emergence of cooperation on Arctic issues and noted that this cooperative approach emerged from a position focused on consolidating Russian sovereignty over its Arctic territories. Russia's approach to Arctic issues rests, accordingly, on a strong preference for the application and development of Russian national laws for the Arctic, with an acceptance of regional arrangements where these complement national functions. Russia thus employs 'subsidiarity' as its guiding principle in its Arctic policies, which provides a set of interlocking national, regional and international positions.

Such an approach ensures that Russia places a strong emphasis on regionalism—notably in the form of the AC—as a means of

Arctic governance. This offers the opportunity to maximize Russian sovereignty, while the pre-eminence of Arctic states in the governance of an 'exceptional' region helps to shield the region from external involvement—except on the terms defined by the regional powers.

However, Zagorski noted that there are some tensions within this position, both in Russia's external relations and in its domestic context. A key issue is the extent to which maximum national advantage can be achieved with regard to the Northern Sea Route (NSR), territorial delimitation and fisheries when regional and international approaches are employed. These potential fault lines in Russian policies could alter the cooperative dynamics in the region.

For Russia, the security dimension stands, to some degree, alone. While the country has not been averse to enhancing security cooperation and finding appropriate platforms for such discussions—although not the North Atlantic Treaty Organization (NATO)—progress has been slow and now risks being frozen as result of the Ukraine crisis.

## The policies of China, Japan and South Korea

In chapter 5, Linda Jakobson and Seong-Hyon Lee discussed the growing interest of China, Japan and South Korea in the Arctic, which is one of most significant recent developments. The economic rise of Asia has transformed the discussion of the Arctic and given deliberations of the Arctic future a whole new context. While climate change threatens key parts of the region's domestic economies, the opening up of the Arctic offers the prospect of new resources for Asia's hungry economies and shorter trade routes to markets in Europe and North America. In all three North East Asian states there has been an increased interest in the Arctic, with more public and private resources flowing into Arctic-related activities. In 2013 they were made permanent observers in the AC—the crowning achievement to date.

As Jakobson and Lee argued, not only did this achievement recognize the interest of these three states in the Arctic, it also acknowledged the legitimate interests of non-Arctic states in future governance mechanisms and structure. China, in particular, would have viewed a rejection of its application for permanent observer status as a hostile signal from the Arctic states to a rising major power.

Yet while similar interests shape the North East Asian states' positions on the Arctic, they also differ significantly in their assessments of the opportunities represented by the Arctic. Thus, while China—and also South Korea, in proportion to its size—is developing capacities and expertise across a range of areas in the Arctic, Japan has been notably more cautious and has even raised the issue of possible security threats to the country as a result of new sea lanes opening to its north.

For North East Asia, the potential opening up of the Arctic places relations with Russia at the centre of their Arctic policies: Russia's position will define their degree of access to some of the key Arctic resources and under what terms they will be able to utilize the NSR. As Russia seeks its own pivot in the Asia-Pacific region, both the pull of Asian demand and Asian investment in the Arctic and the push of deteriorating relations with the transatlantic community, could be a tipping point in the future balance of Arctic relations.

## The centrality of the Arctic Council in Arctic governance

In chapter 6, Svein Vigeland Rottem used the agreement on oil spill preparedness and response as a lens to explore the balance between international and national governance in the Arctic with respect to the core interest of energy. The AC's development as a meaningful international institution over the last decade is one of the most noticeable indications of a new form of Arctic governance.

As Rottem noted, the origins of the AC lie in the thawing of cold war tensions, which freed the region from strategic rivalry and brought other challenges, notably climate change, into focus. In this context, the AC emerged as the organizational embodiment of the new cooperative spirit in the region, and it has made great progress.

The rising significance of the AC as a central governance institution in the Arctic has been reflected in the high-level political engagement that increasingly characterizes ministerial meetings. Likewise, the ambition of key non-Arctic states to participate as observers in the AC underlines the rising international focus on the organization. Here, in particular, it is the work of the AC's working groups, functioning as the 'powerhouse' of the organization, which has been of particular interest. These groups have been central to the emergence of the AC as a leading forum for the production of scientific knowledge about the Arctic, notably environmental issues and climate change.

While the rise of the AC has been underpinned by its role in producing knowledge, its importance as a forum for international cooperation has grown in recent years. This shift has been seen most clearly in the signing of the Agreement on Aeronautical and Maritime Search and Rescue in the Arctic, in May 2011, and the Agreement on Cooperation on Marine Oil Pollution, Preparedness and Response in the Arctic (Oil Spill Agreement), in May 2013. The adoption of these agreements has been widely viewed as reflecting the evolution of the AC from a cooperative group tasked with protecting the environment to an arena capable of negotiating binding international agreements.

The Oil Spill Agreement reflected the Arctic states' core concern over environmental protection but also their keen economic interest in creating the best conditions for exploitation of the Arctic's significant natural resources. The adoption of the agreement marks a significant step in the evolution of the AC, placing it at the centre of governance questions, and the agreement itself reinforces international rules in the region, linking with existing international agreements in this sector.

Yet, as Rottem noted, the Oil Spill Agreement was not drawn up by the AC, it was negotiated by the eight permanent members and the AC constituted the framework for the negotiations. Further, the agreement did not automatically mandate more financial resources or create new institutional capacities. Rather, the focus was on the sovereign obligations of state parties to take steps within the framework of the agreement, and its success will ultimately be determined by the cooperation between the Arctic states.

While the agreement places relatively few shared obligations on the signatories, Rottem argued that it has provided the basis for important practical cooperation in the form of joint exercises; that it improves communication between the Arctic states; and that it has an important symbolic significance. Further, he believes that it highlights the important function of the AC as a mechanism for advancing a cooperative regulatory environment in the Arctic, in the form of 'soft law' instruments, as well as for building consensus through dialogue and the production of scientific information.

Thus, a central argument of the chapter was that the AC operates between international and national levels of governance in the Arctic: its function is to complement rather than replace the Arctic states as the central actors in the region. As such, the AC reinforces rather than weakens sovereignty and it relies on the good relations of the Arctic states in order to function effectively. Perhaps for this reason, Rottem concluded that while the eight member states have signed two binding international agreements in recent years, it is difficult to imagine a further expansion of the AC's operational scope. The AC will remain a decision-shaping body, more than a decision-making one.

## II. Observations and implications for future Arctic governance

New institutions, such as the AC, constitute only one part of the governance architecture that has emerged in the Arctic since the end of the cold war. Informal networks involving non-governmental organizations (NGOs), academics, specialists, businesses and even the military have evolved, focusing on Arctic issues. At the same time, new Arctic institutions have emerged on the foundations of existing multilateral regulatory agreements, notably the United Nations Convention on the Law of the Sea (UNCLOS), which has been the principal framework for resolving contested territorial claims in the region. Arctic governance thus constitutes a sensitive balance between international, regional, national and informal governance structures.

### The Arctic's 'messy' governance structure

Some parliamentarians and even some states have argued that the AC should be established as a fully-fledged international organization because the current architecture of Arctic governance is only a half-completed project.[1] However, as the chapters in this report have highlighted, while the current Arctic governance arrangements may be 'messy', in the words of Alyson Bailes, and leave plenty of scope for further development, they should not be seen as non-cohesive. During a period of significant change, governance in the Arctic has developed in a pragmatic and flexible way across a variety of dimensions and actors, in formal and informal settings, to provide a governance regime that rests on a good degree of political consensus among Arctic actors and thereby provides solid foundations for future development.

### National interests drive Arctic cooperation

Nevertheless, despite the successes since the 1990s, there are very real constraints on regional governance today. A key theme of this report is the 'Arctic spirit of cooperation': a spirit that has made the progress of recent years possible, but which has ultimately rested on a shared interest among the Arctic states to protect and strengthen their sovereignty. The new Arctic governance has, fundamentally, been about building cooperation in order to advance national interests in the region. It has been about creating a set of 'rules of the game' in the region around sovereign claims and the interests of the Arctic states.

The form of cooperative governance that has developed in the Arctic reflects the reality—as in most other regional arrangements around the world—that states remain the central actors in the Arctic, despite progress in building multilateral agreements. In this sense, cooperation has been approached in a pragmatic fashion as the most effective way of achieving national interests rather than as a means of fashioning a supranational governance framework. As Andrei Zagorski noted, 'subsidiarity' has driven Russia's (the largest Arctic state) approach to the Arctic: it has sought to build

---

[1] See e.g. 'Arctic Governance in an Evolving Arctic Region—A Proposal by the Standing Committee of Parliamentarians of the Arctic Region', 10th Conference of Parliamentarians of the Arctic Region: Conference Statement, 5–7 Sep. 2012, <http://www.arcticparl.org/files/arctic-governance-in-an-evolving-arctic-region-2.pdf>, p. 7; and Wilson, P., 'An Arctic Council Treaty? Finland's bold move', eds L. Heininen, H. Exner-Pirot and J. Plouffe, *The 2014 Arctic Yearbook* (Northern Research Forum: Akureyri, Iceland, 2014).

cooperation where national sovereignty could be extended. Moreover, the insistence on respect for sovereign rights by the coastal states remains the basis of Arctic governance at all key levels. It is a prerequisite for developing regional and international instruments governing Arctic activities.

Thus, as Svein Vegeland Rottem argued, the agreements on oil spill response and SAR do not mark the first steps towards the sharing of sovereignty through shared regional governance of the Arctic, but rather demonstrate the firm commitment of the Arctic states to managing the region themselves. Further, according to Rottem, the Oil Spill Agreement strengthens the national responsibility of the Arctic states within their respective sectors. These states not only want to manage the Arctic, but they want to have full control over what happens in their own Arctic territories and waters. For instance, in areas such as offshore oil and gas development, governance mechanisms are primarily based on national regulations, and regional involvement and regulations are, in the words of Rottem, 'loose and go no further than each coastal state has been ready to accept'.

### Arctic states versus Arctic 'outsiders'

The Arctic states—in general terms the eight permanent members of the AC, but in particular the five Arctic coastal states (A5)—wish to maintain their current position of dominance when it comes to deciding on all issues related to Arctic governance. In the 2008 Ilulissat Declaration, the A5 drew a clear line between the Arctic and non-Arctic states in terms of addressing Arctic issues—and this has de facto become the stance of the eight. Understandably, the major non-Arctic states are not comfortable with this stance.[2] Thus, the question of the legitimate interests and rights of 'outsiders'—states that have identified a strong interest in Arctic affairs but are not themselves Arctic states—is pivotal for the future of the region.

---

[2] Young, O. R., 'Listening to the voice of non-Arctic Ocean governance', eds O. R. Young, J. D. Kim and Y. H. Kim, *The Arctic in World Affairs* (Korea Maritime Institute/East-West Center: Seoul/ Honolulu, 2012), p. 282.

Non-Arctic states and intergovernmental bodies have legitimate concerns about how the Arctic will be governed. From a political and economic perspective, the participation of North East Asian states, and in particular China, in Arctic governance is a key issue in building inclusive Arctic governance. Claims by Chinese officials that the Arctic represents a common heritage for humanity (and the subtext that, as an emerging superpower, China should be involved in all areas that it considers important) pose a challenge to the Arctic states' claims to the primacy of regional governance.

At the 2013 Kiruna ministerial meeting, the agreement to admit new permanent observers (including China, Japan and South Korea) indicated progress on this sensitive issue, even if questions surrounding the optimal balance of roles between 'insiders' and 'outsiders' in the Arctic, and their different visions of the future, are likely to remain. Failure to take the first step towards finding agreement on this pivotal issue could have weakened the future prospects of Arctic governance. Yet despite the emphasis on inclusiveness in conjunction with the 2013 decision, none of the Arctic states genuinely wants to relinquish decision-making power to non-Arctic states. The Arctic governance institutions are still designed, in the words of Andrei Zagorski, to 'shield regional decision-making, to the extent possible, from non-regional influences'.

As Oran R. Young points out, permanent observer status does not provide a solution to the issue, that is, the puzzle confronting those concerned with the search for an 'effective means to pay attention to the legitimate concerns of non-Arctic states, without interfering with or disrupting the work of existing cooperative arrangements like the Arctic Council'.[3] If the ice continues to melt at present rates, the growing interests of non-Arctic entities relating to activities such as commercial shipping, oil and gas development, fishing, tourism and environmental protection will have to be addressed. The challenge is to find a way forward that satisfies the essential interests both of Arctic and non-Arctic states. In the long term, the A5 will need to find a way to accommodate the non-Arctic states' desire to be more than merely heard—which currently, as permanent observers, is the most they can hope for.

---

[3] Young (note 2), p. 275.

The fact that permanent observers have committed to accepting the A5's sovereign rights in the Arctic does not necessarily equate to an acceptance of their right to decide on all matters relating to the Arctic. Thus the 'messy' governance that Bailes described is perhaps the best solution one can hope for. Seemingly chaotic conditions do not necessarily equal chaos.

### The rise of 'soft' security cooperation

While recent decades have seen progress within Arctic governance in key socio-economic and environmental issues, the issue of security—particularly 'hard' security—has only moved slowly. In the early 1990s, the drastically improved security environment following the end of the cold war facilitated some progress; and throughout the 1990s and 2000s, positive security relations among the Arctic states continued as an enabler of emerging cooperation. The demilitarization of the Arctic as a result of cold war military infrastructure and weaponry being neglected or withdrawn from the region reinforced the sense that the 'bad old days' of East–West confrontation were far in the past. Instead, the Arctic became the zone for a shared response to the new 'soft' security challenges, initially focusing on environmental issues (in particular climate change) that had local and regional, as well global, implications.

The most serious challenges relate to human and environmental security. Bailes listed food security problems, physical and mental health problems, population movements, and the occurrence of new pests and diseases among the many consequences of climate change affecting the security of individuals and societies in the region. The highly fragile Arctic eco-system also makes it particularly vulnerable: the Arctic would, for example, take much longer to recover from an oil spill or similar disaster than a tropical or semi-tropical zone.[4]

Thus, at first glance, there is a strong case for strict environmental regulations and robust regional environmental governance. However, shipping industry representatives warn that too

---

[4] Welch, D. A., *The Arctic and Geopolitics*, East-Asia Arctic relations: boundary, security and international politics, Paper no. 6 (Centre for International Governance Innovation: Waterloo, Dec. 2013), p. 4.

stringent regulations stipulating, for example, fuels and ship types would decrease the commercial viability of the Arctic sea routes, resulting in less investment, which in turn would have negative economic consequences for the region.[5] Here, as with all things Arctic, a balance needs to be struck.

The melting Arctic ice poses a tangible threat to 'ecospheric security' and this factor has proved crucial in making the Arctic important beyond the immediate Arctic neighbourhood. This hitherto obscure term refers to the threat of fossil fuel emission and the release of greenhouse gases into the ecosphere—in other words, the part of the Earth that supports life.[6] Although China has been one of the most vocal about the global consequences of the melting ice, other non-Arctic states have also emphasized them.[7]

Global sea level rises as a result of the melting polar ice caps threaten the social fabric in numerous societies far from the Arctic. Dislocated populations are potentially politically destabilizing. This in turn could have geopolitical ramifications. Tens—and possibly hundreds—of millions of people will have to abandon their homes as sea levels rise.[8] Seven Asian countries—Bangladesh, China, India, Indonesia, Japan, the Philippines and Viet Nam—are among the top ten countries most vulnerable to rising sea levels.[9]

A global dimension is also strongly evident when one takes stock of the numerous opportunities that the melting Arctic offers. These could enhance energy security and economic security worldwide: people in far-flung places could benefit from new commercial shipping lanes, resource extraction and abundant fishing waters.

---

[5] 'Chinese and Nordic Perspectives on arctic Developments', Comments by industry representatives at a Nordic–China Arctic workshop organized by SIPRI and the China Center for Contemporary World Studies, Beijing, 10 May 2012, <http://www.sipri.org/research/security/arctic/arcticevents/chinese-and-nordic-perspectives-on-arctic-developments>.

[6] Welch (note 4), p. 5. Welch notes that the crucial consideration is the rate of change. The Arctic is relevant because vast quantities of powerful greenhouse gases—carbon dioxide and methane, in particular—are locked up in permafrost.

[7] Welch (note 4), p. 5.

[8] Intergovernmental Panel on Climate Change (IPCC), '10.4.3. Coastal and low lying areas', IPCC Fourth Assessment Report: Climate Change 2007, <http://www.ipcc.ch/publications_and_data/ar4/wg2/en/ch10s10-4-3.html>.

[9] Asia Pacific Foundation of Canada, 'Climate change and the risk of displacement in Asia', Feb. 2014, <http://www.horizons.gc.ca/eng/content/climate-change-and-risk-dis placement-asia>.

Although the commercial viability of all three of these opportunities is still being debated, any or all of them could also increase the risks to human and environmental security. Nevertheless, the compelling opportunities are reason enough for the international community to encourage the Arctic states to seriously consider the global implications of a changing Arctic environment. As the ice melts, encouragement could turn to pressure, depending on how the Arctic states decide to manage the dichotomy between cooperation and competition.

As Bailes noted, many of the challenges arising from global warming cannot be overcome by the region's inhabitants and institutions. Moreover, as the ice melts, new mechanisms to regulate fisheries and shipping in the high seas of the Arctic will need, and must be founded in, international law. However, there is still resistance among some of the Arctic states, notably Russia, to allow the formation of these mechanisms if it is at the expense of losing control of national jurisdiction.

### Arctic regional relations reflect international security

Recognition of the important soft security challenges of the Arctic has spurred a few important legally binding international agreements, such as the SAR Agreement and the Oil Spill Agreement. There is also a budding informal set of networks today, involving coast guards, intelligence organizations, shipping authorities and even militaries. With regard to the main security challenges in the region, Bergh and Klimenko described an emergent circumpolar security architecture, but at the same time they noted that the prospects of a full-blown security community in the Arctic remain slim. This is a reflection of the key role of broader extra-regional security relations between Arctic states in setting the tone for relations within the region: Arctic relations are secondary.

This observation underlines the importance of the wider international security environment for Arctic governance. As noted by Bailes as well as Bergh and Klimenko, the future of Arctic governance is inexorably intertwined with developments in the international order at large and with domestic dynamics within the major Arctic states. The Arctic cannot be singled out as a standalone region without considering the geopolitical complexities and interdependencies that determine interstate relations unrelated to the Arctic or without taking into account the effects of globalization on, for example, technology (to name just one important aspect).

Over the past two decades, the non-threatening climate surrounding traditional 'hard' security questions has been crucial for building cooperation in the Arctic. In such a benign context, even the commissioning of new military and policing capacities has been viewed more as a securitization of the Arctic than a remilitarization of it.[10] Today, there are signs that the basics of Arctic security may be changing. In the immediate post-cold war years, security relations outside the Arctic provided the basis for new cooperative Arctic governance to emerge in the region. But, as a result of the crisis in Ukraine, the recent deterioration in relations between Russia and the transatlantic community has raised questions about the future of the Arctic as a zone for international cooperation.

Although the key Arctic issues remain subject to agreed international processes and hardly cause regional tensions, the fallout from the conflict in Ukraine has affected the region. Security assessments by the transatlantic community have shifted to identifying Russia as a potential threat, notably as a result of its programme of military modernization that includes nuclear and conventional forces based in the Arctic. Within this context, the USA's AC chairmanship (2015–17) is taking place at a crucial moment. The AC's approach to Arctic issues during this period will shape relations in the region in the years ahead. Will the transatlantic community, with the USA in the lead, continue to view the Arctic as a special region where cooperation with Russia can be advanced? In terms of environmental protection, promoting the economic development of the Arctic and continued contacts between the Euro-Atlantic Arctic states (Canada, Denmark, Finland, Norway, Sweden and the USA) and Russia in the realm of soft security (SAR, anti-smuggling and anti-terrorism) seem likely, but the mood in the region risks being overshadowed by the shifting hard security assessments.

---

[10] Wezeman, S. T., *Military Capabilities in the Arctic*, SIPRI Background Paper (SIPRI: Stockholm, Mar. 2012).

Complicating the Arctic dynamics is China's interest in gaining a foothold in lucrative energy projects in the wake of Western sanctions on Russia. Several international giants such as ExxonMobil, Eni and Statoil have been compelled to withdraw from operations in northern Russia, leaving Russian firms in need of financial and technological partners. China has the finances but not the technological know-how, although it is has energetically upgraded its capabilities over the past few years. In this way, the Ukraine crisis could have triggered a tectonic shift in Arctic relations between the transatlantic community and Russia, on the one hand, and Russia and China, on the other hand.[11]

### New challenges for Arctic governance

The Arctic may thus be approaching a turning point in regard to the form of governance that develops in the future. There remains a broad consensus among the Arctic states on the importance of resolving the region's challenges in a cooperative manner and on the basis of international law. The idea of stewardship of the Arctic continues to be a shared commitment among the Arctic states and the foundation for advancing sustainable development and environmental protection as a collective enterprise. Further, UNCLOS is recognized as the basis for the equitable delimitation of the region, even if the USA has yet to ratify it. However, the agreement on the underpinnings of Arctic political and security relations is being challenged by the complex geopolitical developments surrounding the conflict in Ukraine and the fallout from it that has spread across the wider region.

While security issues alone will not determine the future of Arctic governance, they will play a central role in shaping the landscape of Arctic relations and establishing the boundaries of cooperation. With the 'spirit of Arctic cooperation' facing an unprecedented challenge, finding ways to manage security relations in the region is likely to become a more urgent challenge, alongside the soft security agenda that has driven the creation of the existing forms of Arctic governance.

---

[11] Trenin, D., *From Greater Europe to Greater Asia? The Sino-Russian Entente*, Carnegie Moscow Center Paper (Carnegie Moscow Center: Moscow, Apr. 2015).

# About the authors

**Alyson J. K. Bailes** (United Kingdom) spent most of her career as a British diplomat, working in posts from Budapest to Beijing, before becoming Director of SIPRI from 2002–2007. Since 2007 she has been an Adjunct Professor at the University of Iceland, specializing in security studies, and has also taught at the College of Europe in Bruges. Her current research interests include Nordic cooperation and 'small state' studies as well as Arctic affairs. Her publications include *Small States and International Security: Europe and Beyond* (Routledge, 2014, co-editor).

**Kristofer Bergh** (Sweden) is a former Researcher with the SIPRI Armed Conflict and Conflict Management Programme. He has also worked on European Security and Global Health projects at SIPRI. He holds a Master's degree in Peace and Conflict Studies from Uppsala University and his main research interests include Arctic security and US and Canadian Arctic policy.

**Linda Jakobson** (Finland) is a Sydney-based independent researcher specializing in Chinese foreign policy and East Asian security. She is also the founding director of China Matters, a public policy initiative focused on Australia–China relations. Before moving to Australia in 2011, Jakobson lived and worked in China for 22 years and published six books on Chinese and Asian society and politics. She has also published extensively on Chinese foreign policy, maritime security in Indo-Pacific Asia and China's Arctic aspirations. Her last position in Beijing was Director of SIPRI's China and Global Security Programme (2009–11).

**Ekaterina Klimenko** (Kyrgyzstan) is a SIPRI Researcher, currently conducting research as part of the Arctic Futures Project and the Conflict and Peacebuilding in the Caucasus Project. She holds a Master's degree in International and European Security from the University of Geneva and is a PhD candidate at the University of Helsinki. Previously, she worked as a Research and Training Assistant at the Organization for Security and Co-operation in

Europe (OSCE) Academy in Bishkek and as an intern at the OSCE Secretariat.

**Dr Seong-Hyon Lee** (South Korea) is an Assistant Professor at Kyushu University in Japan. He is also a Senior Research Fellow at the Peking University Center for Korean Studies, a Salzburg Global Fellow and a James A. Kelly Fellow of the Pacific Forum Center for Strategic and International Studies (CSIS). A native of Seoul, he was a Pantech Fellow at the Shorenstein Asia–Pacific Research Center of Stanford University (2013–14). His research interests include China–North Korea relations, Chinese media and foreign policy, North Korea, and China–South Korea relations.

**Dr Neil Melvin** (United Kingdom) is a Senior Researcher at SIPRI, where he specializes in the study of conflict and conflict management, with a particular regional focus on Eurasia. Prior to joining SIPRI, he held senior adviser positions in the Energy Charter Secretariat and the Organization for Security and Co-operation in Europe (OSCE). He has worked at a variety of leading policy institutes in Europe and published widely on issues of conflict.

**Dr Svein Vigeland Rottem** (Norway) is a Senior Researcher and the Director of the Russia and Polar Programme at the Fridtjof Nansen Institute (FNI). His PhD was on the Norwegian defence establishment's encounter with new post-cold war realities, emphasizing, among other things, security and safety in Arctic waters. In recent years, his main research focus has been on Arctic governance, safety and security issues in the Arctic, and the Arctic Council, on which he has published widely.

**Dr Andrei Zagorski** (Russia) is Head of the Department for Arms Control and Conflict Resolution at the Institute of World Economy and International Relations (IMEMO) of the Russian Academy of Sciences. He is a leading expert in European security and Russia's relations with multilateral European organizations, such as the Organization for Security and Co-operation in Europe (OSCE), the European Union (EU) and the North Atlantic Treaty Organization (NATO). His area of expertise also includes Russian foreign and security policy, post-Soviet studies, arms control and Arctic studies. Previously, he participated as an expert in Soviet delegations at several meetings of the Conference on Security and Cooperation in Europe (CSCE). He has been active in research, consulting and teaching, including serving as Vice-Rector (Research Director) at the Moscow State Institute of International Relations (1992–99).

# Index

Abe, Shinzo 134
Ahlenius, Hugo 3, 85, 97, 112
Alaska:
  federal vs state policies 46, 74
  new commercial activity 25
  oil and gas 32, 47, 160
  Open Skies Treaty and 21
  shipping 47
Antarctic 16, 93
Antarctic Treaty 29
Arctic Circle 63, 73
Arctic Climate Impact Assessment (ACIA, 2004) 152–53, 159
Arctic Coast Guard Forum 39–40, 104
Arctic Council:
  Arctic states vs outsiders 183–85
  biodiversity and 171
  chairs 42–43, 44, 48, 65–66, 101, 155, 188
  China and 39, 111, 114–15, 123–24
  civil protection and 33
  climate change and 28, 152–53, 172
  consensus base 74
  cooperation and participation 149, 152–58
  decision-making structure 149–52
  differences of opinion 155–56
  drivers 176
  Environmental Protection Strategy 28
  EPPR working group 164
  fishing regulation 171
  future 170–72
  Gorbachev on 67
  Guidelines 169
  IMO and 170
  indigenous communities and 156–57
  membership 4, 41n2, 158
  nature 4
  Nordic countries and 60
  North East Asia and 145–46, 184
  China 39, 111, 114–15, 123–24, 156, 175, 184
  Japan 39, 111, 131, 132, 156, 183
  South Korea 111, 114, 141–42
  observer status 5–6, 39, 65, 67, 73, 111, 114–15, 156–57, 158, 175, 184
  Oil Spill Agreement *see* Oil Spill Agreement (2013)
  origins 105, 148–49, 152–53, 174, 179
  permanent participants 36, 156–57, 158
  permanent secretariat 147
  platform for agreements 4
  political renaissance 157
  research 150, 161
  role 7, 75, 145, 147, 173, 179–81
  Russia and 57, 59, 103, 104–107
  SAR Agreement *see* SAR agreement
  security and 20, 37, 66, 101, 103
  soft security 177
  shipping code 30
  societal security 35
  soft law 169–70
  sovereignty and 107
  status 12
  Sweden and 48, 65–66
  UNCLOS and 155–56
  United States and 48, 101
  working groups 146, 150–52, 166, 180
Arctic Economic Council 31–32, 44
Arctic Environmental Protection Strategy (AEPS) 149, 174
Arctic Five 105, 108, 109, 145, 146, 154, 183–85
Arctic Frontiers Conference (2013) 147
Arctic governance:
  Arctic Council *see* Arctic Council
  Arctic states vs outsiders 183–85
  future 181–89

globalization and 5–6
key themes 4–10, 174–81
legally binding mechanisms
  166–68
messy structure 38, 175, 181–82, 185
national interests driving
  cooperation
  182–83
non-binding mechanisms 169–70
public vs private actors 6–7
security *see* security
sovereignty *see* sovereignty
sustainability 7–8
*see also* specific conventions and
  organizations
Arctic region: map 3
Arctic Science Summit Week (2015)
  128
Arctic Security Forces Roundtable
  (ASFR) 33, 69, 101, 103
Australia: China and 119

Bailes, Alyson 11, 13–40, 41, 73,
  99–100, 102, 174–75, 182, 185, 187
Ban Ki-moon 138
Bangladesh: climate change and 186
Barents 2020 initiative 170
Barents Euro-Arctic Council (BEAC)
  20, 28, 35, 36, 105
Beaufort Sea 42
Bergh, Kristofer 1–12, 41–75, 102,
  176–77, 187
biodiversity 171
Biodiversity Convention (1992) 78
BP 7, 32
Brazil: Arctic interests 145

Canada:
  Arctic Council and
    chairmanship 42, 44
    membership 4
    origins 149, 153
    permanent observer issue 114,
      156
  Arctic Economic Council and 31,
    44

Arctic Foreign Policy 43
Arctic littoral state 16, 154
China and 124, 126
continental shelf claims 44–45, 61,
  62, 83
cooperation objective 149
Denmark dispute 4–5
energy security 26–27
environmental policy 44, 91
EU Arctic role and 62–63
fracking 26–27
Japan and 135
NATO and 22, 45, 100
navigation rights 87–88
new commercial activity 25
Northern Strategy 43
oil exploration 159
Oil Spill Agreement exercises 163
Open Skies Treaty and 21
Russia and 52
security 18
  cooperation 68
  Iceland agreement 23
  joint exercises 68
  Operation Nanook 67
  policy 17, 42–45, 73
  procurement 16
South Korea and 143
Ukraine crisis and 44, 69–70
UNCLOS and 91
United States and
  cooperation 22, 100
  Northwest Passage 176
Chealsey ministerial meeting (2010)
  105
China:
  AC observer 39, 111, 114–15, 123–24,
    124, 156, 175, 184
  Arctic actors 117–23
    commercial entities 120–23
    COSCO 120–21
    government entities 117–19
    Polar Research Institute of
      China 119, 120
    research institutions 119
    resource companies 122–23
    shipping companies 120–22

State Oceanic Administration 117, 119
Arctic interests 143, 184
  actors 117–23
  drivers 115–17, 144
  economic interests 21, 25
  energy 189
  fishing 96, 114
  Japan and 133
  outlook 146
  overview 111–27
  policies 123–24, 178–79
  resources 122–23, 125–26
  scientific research 115–17, 119, 130
  trade 120–23
Arctic littoral states and 124–27
Canada and 124, 126
climate change and 28, 114, 117, 186
Denmark and 126
EU–China Summit (2012) 121
Finland and 119
Iceland and 63, 126–27
Japan and 133, 135
maritime disputes 135
Norway and 126
private and public spheres 7
Russia and 37, 72, 124–26
  energy 125–26, 189
  outlook 146
  security 134, 143–44
Ukraine crisis and 124
United States and 124, 126
China National Petroleum Corporation 125–26
Chuangli Group 122
Chukchi Sea 47
climate change:
  anthropogenic factors 28
  Arctic Council and 28, 152–53, 172
  China and 28, 114, 117, 186
  effect 1
  melting sea ice 24–25
  migration and 36
  mitigation 25, 172
  negotiations 152
  new opportunities 11, 24
  research in Arctic 11, 111, 116–17
  security and 14, 186
  South Korea and 139
Clinton, Hillary 46, 70, 145, 150, 154
cold war 1, 2, 16, 22, 67, 88, 100, 175, 185
commercial interests:
  economic security 25–26
  public actors and 6–7
  *see also* individual countries
Commission on the Limits of the Continental Shelf (CLCS) 61–62, 72, 81–87
common heritage of mankind 8, 82, 84, 86, 108, 184
continental shelf:
  Canada 44–45, 61, 62, 83
  Denmark 61–62, 83
  Greenland 61–63, 83
  Norway 83, 85
  regulation 167
  Russia 55–56, 72, 81–87
  UNCLOS 4, 55–56, 72, 81–87, 167
  United States 83, 87
COSCO 120–21
customary law 78, 80, 86, 87, 107, 165

Denmark:
  Arctic Council and 4, 154, 155
  Arctic littoral state 16, 154
  Canadian dispute 4–5
  China and 126
  continental shelf claims 61–62, 83
  Faroe Islands 4, 26, 32, 35, 36, 64
  OSPAR Convention and 168
  Polar Code and 93
  Russia and 52
  security
    Iceland agreement 23
    NATO strategy and 22
    Operation Nanook 67
    policy 59–62, 74
    procurement 17
  sovereignty and 37
  *see also* Greenland

energy:
    Canada 26–27
    China–Russia relations and 125–26, 189
    European Union 29
    Iceland 26
    Japan 127, 129, 134
    security 26–29
    *see also* oil and gas
environment:
    Arctic Environmental Protection Strategy (AEPS) 149
    environmental impact assessments 31
    environmental security 23–32, 185–87
    mining and 31–32
    outsiders and 184
    shipping 29–30
    *see also* climate change
Eriksen Søreide, Ine Marie 70
European Union:
    AC observer status 156
    Arctic role 62–63, 74
    climate change and 28
    energy security 29
    EU–China Summit (2012) 121
    modified sovereignty 16, 37
    Nordic members 59
    Northern Dimension 20, 28
    Ukraine crisis and 70
exceptionalism: international law and 8–9
Exxon 27–28, 71

Faroe Islands 4
    Arctic strategy 36
    civil protection 32
    emigration 36
    Icelandic cooperation 64
    oil and gas 26
    West Nordic Council membership 35
Finland:
    Arctic cooperation report (2015) 65
    Arctic Council and 4, 154
    Arctic Five and 145
    biodiversity and 171
    China and 119
    cooperation objective 149
    Iceland and 64
    polar research 62, 74
    Rovaniemi process 149
    security
        joint exercises 23, 68
        NATO 23, 100
        policy 59–61, 62–63, 74
        Swedish cooperation 100
    sovereignty and 37
Fish Stock Agreement 95
fishing:
    2015 Declaration 98
    Arctic Council and 171
    China 96, 114
    Iceland 30, 63, 64
    importance 30–31
    international law 8–9, 95
    map 97
    Norway 161, 171
    regulation 171, 187
    Russia 95–98, 107, 171
fracking 26–27
France: ASFRs and 101

Gazprom 7, 20, 50–51, 170
Germany: ASFRs and 101
global commons 80
Global Green Growth Institute 139
global warming *see* climate change
globalization 5–6, 14, 38, 108, 124, 188
Goose Bay ministerial meeting (2012) 33
Göranson, Sverker 66
Gorbachev, Mikhail 66–67, 148
Green Climate Fund 139
Greenland 4
    civil protection 32
    continental shelf 61–62, 83
    cruise traffic 160–61
    fishing 30
    Icelandic cooperation 64
    independence debate 61, 74

migrants 36
new commercial activity 25
oil and gas 26, 159
security 74
West Nordic Council membership 35
*see also* Denmark
Greenpeace 6, 19–20, 163
Grimsson, Ólafur Ragnar 63

Haga process 34
Hans Island 4, 62
Harper, Stephen 44, 114, 143
Hashimoto Yasuaki 130
Helsinki ministerial meeting (2011) 34
human trafficking 9
Hunchun Chuangli Haiyun Logistics 121–22
Hyundai 122, 136

Iceland:
  2008 crash 25n41
  Arctic Circle and 63
  Arctic Council and 4, 154
  Arctic Five and 145
  Arctic littoral state 16, 63
  Arctic strategy 36
  China and 63, 126–27
  civil protection 32
  Finland and 64
  fishing 30, 63, 64
  security
    air surveillance 64
    bilateral defence agreements 22–23
    cooperation 22–23, 64
    energy security 26
    NATO membership 21, 64
    Norwegian cooperation 23, 64
    policy 59–61, 63–64, 74
    procurement 16
  Sweden and 64
  UNCLOS and 16
  United Kingdom and 64
  United States and 64

West Nordic Council membership 35
Ilulissat Declaration (2008) 75, 80, 93, 145, 183
Ilulissat ministerial meeting (2008) 62, 105, 155–56
India:
  AC observer 39, 114, 156
  Arctic interests 145
  climate change and 186
indigenous communities 6, 19, 24, 28, 35, 36, 41, 61, 145, 151, 156–57, 173
Indonesia: climate change and 186
International Council for the Exploration of the Sea (ICES) 95
international customary law 78, 80, 86, 87, 107, 165
International Institute of Strategic Studies 16
International Maritime Organization (IMO) 30, 88, 93, 106, 108, 162, 165, 167–68, 170
International Mercury Convention 152
International Organization for Standardization (ISO) 170
International Seabed Authority 8, 82
Iqaluit ministerial meeting (2015) 156
Italy: AC observer 156

Jacoby, Charles 68
Jakobson, Linda 11, 90, 96, 111–46, 174–89
Japan:
  AC observer 39, 111, 131, 132, 156, 184
  Arctic interests 111–13
    actors 131–32
    caution 143–44
    commercial actors 129–30
    drivers 127–30
    NSR 127, 129, 132–33
    overview 127–35
    policies 132–33, 178
    scientific research 131, 132
    shipping 127, 128, 129, 132–33

strategic drivers 130, 133
Arctic littoral states and 134–35
Arctic Task Force 128, 131
Canada and 135
China and 133, 135
climate change and 28, 186
energy needs 127, 129, 134
Institute of International Affairs 131–32
JOGMEG 131
Norway and 134
nuclear pollution cooperation 28
Ocean Policy Headquarters 131
Ocean Policy Research Foundation 131
oil and gas 127
Russia and 133, 134
   Kurile Islands 127
security 127
United States and 133

Kerry, John 46, 72, 150, 154
Kiruna Declaration (2013) 93, 109
Kiruna ministerial meeting (2013) 5, 33, 65, 72–73, 147, 150, 154, 156, 158, 184
Klimenko, Ekaterina 11, 41–75, 102, 176–77, 187
Korea, Democratic People's Republic of (DPRK) 121–22, 133
Korea, Republic of *see* South Korea
Korolev, Vladimir 53
Kotani, Tetsuo 129, 135
Kurile Islands 127

Lavrov, Sergei 52, 56, 72, 150
law of the sea:
   MARPOL *see* MARPOL
   Russia and 77–81
   SOLAS *see* SOLAS
   sovereignty and 104
   UNCLOS *see* UNCLOS
Lee Myung-bak 138
Lee, Seong-Hyon 11, 90, 96, 111–46, 178–79
Lee Seong-woo 139

Li Keqiang 119
Liu Xiaobo 126
Lloyd's of London 30
LNG 125–26, 129, 134

Makarov, Nikolai 68
MARPOL 8, 30, 78, 92, 93, 167–68
Medvedev, Dmitry 52
Melvin, Neil 1–12, 174–89
migration 36
mining 24, 29, 31–32, 109
   *see also* oil and gas
Mitsui O.S.K. Lines 134
Mizincev, Mikheil 54
Murmansk Initiative (1987) 66–67, 148

NATO:
   Arctic littoral state members 21, 99
   Arctic region and 9, 10, 13, 22–23, 38, 73, 74
   Canadian Arctic strategy and 22, 45
   exercises in Nordic countries 23, 60–61
   Iceland and 21, 64
   Norway and 22, 64–65, 74
   Partnership for Peace 67, 100
   Russia and 17–18, 21, 22, 178
      energy security 27
      NATO–Russia Council 101, 103
      threat assessment 52, 58, 73
   SAR exercises 34
   USA–Canada cooperation 22, 100
   US strategy and 22, 46
Natynczyk, Walter 68
navigation *see* shipping
Netherlands: ASFRs and 101
New Siberian Islands 16, 54
NGOs 6, 9, 19–20, 62, 181
Nordic Declaration of Solidarity (2011) 59–60
Nordic Defence Cooperation (NORDEFCO) 23, 60, 68
North American Aerospace Defense Command (NORAD) 42, 68

North Atlantic Coast Guard Forum 101
North Atlantic Salmon Conservation Organization (NASCO) 95, 96
North East Atlantic Fisheries Commission (NEAFC) 31, 95–97
North Korea: Rajin Port 121–22, 133
North Pacific Coast Guard Forum 101
Northern Sea Route (NSR):
  China and 114, 120, 121, 125
  commercial potential 6, 29
  growing traffic 160
  Japan and 127, 129, 132–33
  North East Asia and 111, 143, 179
  Northwest Passage and 111, 113
  Russia and 178
    China and 125
    development 50, 178
    regulation 29, 94
    sovereignty focus 90
    strategy 49, 88–92
  South Korea and 135, 136, 137–38, 141, 142
Northwest Passage 42, 111, 113, 160, 176
Norway:
  Arctic cooperation report (2015) 65
  Arctic Council and
    chairmanship 155
    membership 4
    Oil Spill Agreement 161, 162, 172
  Arctic littoral state 16, 154
  Arctic policy 165
  Barents Sea 42, 57, 64, 65, 83
  China and 126
  continental shelf 83, 85
  EEZ: map 85
  energy security 27
  fishing 161, 171
  Iceland and
    air surveillance 64
    defence agreement 23
  Japan and 134
  oil and gas 26, 159
  Oil Spill Agreement and 161, 162, 172
  OSPAR Convention and 168
  Polar Code and 93
  Russia and 100
    2010 Treaty 42, 57, 64, 67, 83
    cooperation 164
    fishing arrangements 171
  security 16–17
    cooperation 69
    Iceland defence agreement 23
    joint exercises 68, 100
    NATO 22, 64–65, 74
    Partnership for Peace 67
    policy 59–61, 64–65, 74
  societal security 34
  South Korea and 142
  sovereignty issues 64
  Ukraine conflict and 70
  UNCLOS and 165
Novatek 70, 71, 125–26
Nuuk ministerial meeting (2011) 33, 145, 154

Obama, Barack 46, 47–48
oil and gas:
  Alaska 32, 47, 160
  Canada 159
  exploration 159–61
  Faroe Islands 26
  Greenland 26, 159
  Japan 127
  map 112
  Norway 26, 159
  outsiders 184
  pollution 31–32
  private actors 7
  Russia 6, 26, 27, 49, 71, 159
  soft law 170
  undiscovered resources 159
  *see also* Oil Spill Agreement (2013)
Oil Spill Agreement (2013):
  AC mandate 169–70
  contents 162–63
  international framework 162, 165
  joint exercises 163
  need 158–61
  negotiations 148, 180
  Norway and 162, 172

objectives 162
origins 147
Russia and 155, 162
scope 163–66, 169
significance 103, 106, 148, 153, 180, 183
soft law 187
sovereignty 4, 180
United States and 154
value 165–66, 173
Open Skies Treaty 21
Organization for Economic Co-operation and Development (OECD) 136
Organization for Security and Cooperation in Europe (OSCE) 20–21
OSPAR Convention (1992) 168
Ottawa Declaration (1996) 174

Park Geun-hye 141, 142, 143
Park Jin-hee 137
Particularly Sensitive Sea Areas (PSSAs) 30
Partnership for Peace 67, 100
Patrushev, Nikolai 51–52
Peng Liyuan 119
Philippines: climate change and 186
Polar Code 92–95, 106, 108, 165, 167–68
Pomor 57, 68
precautionary principles 171
private actors: public actors and 6–7
protected areas 171
Putin, President Vladimir 48, 50, 55, 57, 70, 71, 125, 130, 142

Qu Tanzhou 115, 124

realism 16
regional fisheries management organizations (RFMOs) 95–98
research:
    Arctic Council and 150, 176
    Arctic Science Summit Week 128
    China 115–17, 119, 130

climate change 11, 111, 116–17
Finland 62, 74
fishing 171
Japan 131, 132
oil and gas extraction and 161
South Korea 130, 140–41, 142, 143
Sweden 62, 74
Rosneft 7, 51, 70, 71, 170
Rottem, Svein Vigeland 12, 147–73, 179–81, 183
Rovaniemi process 149
Russia:
    Arctic Council and
        agreements 59, 155
        membership 4
        permanent observers 114, 156
        security issues 103
        strategy 57, 104–107, 154–55
    Arctic governance 11, 76–110
    Arctic littoral state 16, 154
    Barents Sea 42, 57, 64, 65, 83
    BEAC and 105
    Canada and 52
    China and 37, 72, 124–26
        energy 125–26, 189
        outlook 146
        security 134, 143–44
    continental shelf claims 55–56, 72, 81–87
    environmental jurisdiction 88–92, 107–108
    European Union and 156
    external actors and 76–77
    Japan and 133, 134
        Kurile Islands 127
    law of the sea and 77–81
    Maritime Doctrine (2015) 49, 58
    Military Doctrine (2014) 58
    NATO and 17–18, 21, 22, 178
        energy security 27
        NATO–Russia Council 101, 103
        threat assessment 52, 58, 73
    navigation rights 87–95
        fisheries 95–98, 107, 171
        NSR 88–92, 94
        Polar Code 92–95, 106, 108
    Norway and 100

# INDEX 201

2010 Treaty 42, 57, 64, 65, 83
  cooperation 164
  fishing arrangements 171
NSR *see* Northern Sea Route (NSR)
nuclear pollution cooperation 28
oil and gas 6, 26, 27, 49, 71, 159
Oil Spill Agreement and 155, 162
private and public spheres 7
sectorial approach 80–81
security 16
  ASFRs 103
  changing policy 55–59
  cooperation 55–58, 68, 69, 103–104
  fragmented architecture 99–101
  Japan and 134
  joint exercises 68, 100
  military capacities 52–55, 130
  Nordic countries and 60
  policy 48–59, 73, 102–104, 177–78
  posture 17–18
  procurement 13
  rhetoric 58, 59
  SAR centres 54–55
  sovereignty protection 51, 102, 177–78
  strategic importance 48–50
  threat assessment 50–52
  Ukraine effect 58, 104
shipping 29
Siberia 25, 27
South Korea and 142
sovereignty focus 51, 76, 79, 90, 102, 108, 177–78, 187
subsidiarity 76–77, 107–10, 177–78, 182–83
Sweden and 66
territorial ambitions 70
Transport Strategy 49
Ukraine and *see* Ukraine crisis
UNCLOS and 55–56, 59, 77–81, 84, 91, 107
United States and
  ballistic missile programme 18
  continental shelf 87
  diplomatic model 37
  security 144
  strategic balance 52
  Ukraine crisis 72–73, 100–101
  Western relations 177

Samsung 136
SAR Agreement (2011):
  negotiations 4, 67
  origins 147
  Russia and 103, 106, 155
  significance 100, 153, 180, 183
  soft law 187
  United States and 101, 154
Schissler, Mark 69
sealing 63, 156
security:
  actors 175
  Arctic definition 174–76
  civil emergencies 32–34, 175
  disasters 32–34
  economic security 23, 25–26
  emerging circumpolar architecture 66–73
  energy security 26–29, 175
  environmental security 23–32, 185–87
  fragmented architecture 99–101
  governance and multidimensional security 37–40
  hard security 15–23, 66, 175, 177, 188
  human security 34–37, 175
  institutions 20–21
  issues 9–10, 13–40
  joint military exercises 67–68
  national approaches
    diversity 11, 41–75, 176–77
    Nordic countries 59–66
    North America 42–48
    Russia *see* Russia
    *see also* individual countries
  realist politics 16
  regional and international relations 187–89
  societal security 34–37, 175

soft security 9–10, 13, 177, 185–87, 187
sovereignty and 18–19, 37–40, 73, 75, 175–78
Ukraine crisis *see* Ukraine crisis
violent non-state actors 19–20
Severnaya Zemlya 16
Shell 7, 32, 47
shipping:
  Chinese companies 120–22
  environment and 186
  expansion 24, 29–30, 160–61
  international law 8–9
  Japan 127, 128, 129, 132–33
  non-Arctic states 90
  North East Asia 143
  NSR 88–92, 90, 94, 125, 127
  outsiders 184
  Polar Code 165
  regulation 187
  Russian navigation rights 87–95
  South Korea 122, 137–38
shipping firms 7
Singapore:
  AC observer 114, 156
  shipping hub 133
SOLAS 8, 78, 92, 93, 94
Song Gi-seon 139
South Korea:
  AC observer 111, 114, 141, 141–42, 156, 184
  Arctic Comprehensive Initiative 135–36
  Arctic interests
    actors 139–41
    commercial interests 136–38
    drivers 136–39, 144
    government actors 139–40
    international status 138–39
    NSR 135, 136, 137–38, 141, 142
    overview 111–13, 135–43
    policies 141–42, 178, 179
    research 130, 140–41, 142, 143
    resources 136–37
    shipping 122, 137–38
  Arctic littoral states and 142–43
  Busan 132, 138, 141
  Canada and 143
  climate change and 139
  economic performance 136, 137
  Korea Polar Research Institute (KOPRI) 139–41
  Norway and 142
  OECD membership 136
  Russia and 142
  shipbuilding 141, 142
  United States and 142
sovereignty:
  Arctic Council and 107
  Arctic littoral states 166
  biodiversity and 171
  delimitation process 73
  integration and 4–5
  Oil Spill Agreement and 180
  regional frameworks and 104
  Russian focus 51, 76, 79, 90, 102, 108, 177–78, 187
  security and 18–19, 37–40, 73, 75, 175–78
Soviet Union:
  Murmansk Initiative (1987) 66–67, 148
  NSR 88, 89
  *see also* Russia
Statoil 7, 170
Stockholm Convention on Persistent Organic Pollutants (2001) 152
Stoltenberg Report (2009) 23, 59
subsidiarity 76–77, 107–10, 177–78, 182–83
sustainability: multilateralism and 7–8
Sveinsson, Gunnar Bragi 126–27
Sweden:
  Arctic cooperation report (2015) 65
  Arctic Council and 154
    chairmanship 48, 65–66
    membership 4
  Arctic Five and 145, 154
  Arctic identity 65
  biodiversity and 171
  Iceland and 64
  polar research 62, 74
  Russia and 66

INDEX 203

security
   Finnish cooperation 100
   joint exercises 68
   NATO exercises 23, 61
   policy 59–61, 65–66, 74
   societal security 34
   sovereignty and 37

Toriumi, Shigeki 127
Total 71, 170
tourism 30, 161, 184

Ukraine crisis:
   Arctic Council and 156
   Arctic security and 39–40, 176
   Canadian response 44, 69–70
   Chinese response 124
   effect 2, 10, 11, 18, 23, 189
   EU response 70
   international cooperation and 188
   international sanctions 69–71
   Norwegian response 70
   responses 44, 69–73, 75
   Russian security and 58, 104
   US response 70, 72–73, 98, 100–101
UNCLOS:
   Arctic Council and 155–56
   Arctic littoral states 16
   Canada and 44
   continental shelf regulation 4, 55–56, 72, 81–87, 167
   cooperation and 167
   environmental jurisdiction 88, 92, 94, 107–108
   Fish Stock Agreement 95
   framework 8
   global commons 80
   military navigation 94
   navigation rights 87, 88, 89, 91
   primacy 165, 166–67, 181, 189
   PSSAs 30
   Russia and 55–56, 59, 77–81, 84, 91, 107
   United States and 8, 47–48, 83, 90, 189

United Kingdom:
   Arctic policy 28–29
   ASFRs and 101
   Iceland and 23, 64
United Nations Development Programme (UNDP) 35
United States:
   AC membership 4
   Arctic Council and 45, 48, 154
      agreements 154
      chairmanship 42–43, 101, 188
   Arctic littoral state 16, 154
   Arctic Strategy 46
   BP Deep Horizon explosion 32
   Canada and: Northwest Passage 176
   China and 124, 126
   continental shelf claims 83
   cooperation objective 149
   energy security 26
   fracking 26–27
   Iceland and air surveillance 64
   Japan and 133
   National Strategy for the Arctic Region 46
   NATO and
      Canadian cooperation 22, 100
      strategy 22, 46
   Navy Arctic Roadmap 47
   oil exploration 159, 160
   Oil Spill Agreement and 162, 163
   Polar Code and 93
   Russia and
      ballistic missile programme 18
      continental shelf 87
      diplomatic model 37
      joint exercises 100
      security 144
      strategic balance 52
      Ukraine crisis 72–73, 100–101
   security
      ASFRs 33, 69, 101, 103
      cooperation 69
      joint exercises 68, 100
      Operation Nanook 67
      policy 42–43, 45–48, 74

procurement 17
USEUCOM 42, 69
USNORTHCOM 42
South Korea and 142
Soviet Union and NSR 88
Ukraine conflict and 70, 98, 100–101
UNCLOS and 8, 47–48, 83, 90, 189
*see also* Alaska

Vasiliev, Anton 56, 142
Viet Nam: climate change and 186

West Nordic Council 35
World Wildlife Fund for Nature (WWF) 163

Xi Jinping 119
Xu Minjie 120

Yang, Huigen 120
Young, Oran 145–46, 184

Zagorksi, Andrei 11, 76–110, 177–78, 182–83, 184
Zhang Gaoli 118–19